VOL.16
JOURNAL OF ARMORED
ASSAULT
&
HELIBORNE WARFARE
AF323938

CONCORD
PUBLICATIONS COMPANY

VOL.16
JOURNAL OF ARMORED
ASSAULT
& HELIBORNE WARFARE
7816
"Gators" in Fallujah
"European Challenge" for the Bundeswehr
Leopards against Leopards
Piranha of the Spanish Marines

"Gators" in Fallujah
AAVP7A1 RAM/RS of 2nd Assault Amphibian Battalion
Carl Schulze & Yves Debay
STOP

"European Challenge" for the Bundeswehr
Carl Schulze

Leopards against Leopards
German-Polish Tank Maneuver "Salchauer Schild"
0282
Walter Böhm

BDMZ-III
The Mechanized Landing Battalion of the Spanish Marines
Juan Pablo Lasterra

"Gators" in Fallujah
AAVP7A1 RAM/RS of 2nd Assault Amphibian Battalion

Carl Schulze & Yves Debay

A view of the right hand side of the AAVP7A1. The UGWS (Up-Gunned Weapon Station) is offset to the right of the vehicle's centerline. The vehicle has a gross weight of 21,027kg, but when loaded up with its twenty-one Marine passengers, the combat weight climbs to 25,761kg. (Yves Debay)

The noise of heavy armored vehicles can be heard, but in fact something shaped like a ship traveling on tracks appears out of the dust clouds. On top of this "thing", a heavily armed turret is visible. The first vehicle is followed by a second, and then a third appears. The vehicles thunder along at high speed through the Iraqi desert in the direction of the road visible in the distance. It is mid-December 2005, and AAVP7A1 (Amphibious Assault Vehicle Personnel 7A1) RAM/RS vehicles of A Company, 2nd Assault Amphibian Battalion, conduct an IED sweep along one of the main roads connecting Fallujah with Baghdad. With its armored vehicles the company supports Regimental Combat Team 8, a force that is responsible for security in the Fallujah region. Basically this means that the company provides the RCT's infantry units (which can only field wheeled and lightly armored vehicles) with an armor-protected, tracked transport capability as and when required. In addition, the vehicles of the company support the infantry with heavy fire support by using their main armaments during operations.

AAVP7A1 RAM/RS service history

The AAVP7A1 RAM/RS is an amphibious, armored, fully tracked landing vehicle. The vehicle was designed to carry troops during amphibious operations from ship to shore, through rough water and the surf zone, and over beaches that offer little cover. Originally the Amphibious Assault Vehicle Personnel 7A1 RAM/RS entered service as the "Landing Vehicle Personnel Tracked Model 7" in 1971. It had replaced the "Landing Vehicle Tracked Personnel Model 5" of the United States Marine Corps' ageing fleet. Like the earlier amphibious vehicle, the LVTP7 was a member of the family of vehicles that also included the LVTR7 Recovery vehicle and the LVTC7 Command Post vehicle. It was estimated that the LVTP7 had the durability for a service life of ten years. But the vehicle proved much more capable and, due to this, the USMC started a Service Life Extension Program (SLEP) in 1977. Upgraded vehicles were designated as LVTP7A1 and featured a more powerful engine, passive night vision equipment, an improved suspension and hydraulic system, as well as a modified troop compartment ventilation system.

Between 1987 and 1990, the USMC replaced the original machine gun turret of the LVTP7A1 with a new turret, the so-called "Up-gunned Weapon Station". The one-man turret featured a 12.7mm M48 Turret-

An overall view of the sandy motor pool where A Company of 2nd Assault Amphibian Battalion keeps its vehicles. The trim vane used during amphibious operations lies on the sandy desert floor in front of the nearest AAV. In view are sand-colored M1A1 tanks belonging to the 2nd Tank Battalion, also quartered at this same facility. (Yves Debay)

AAVP7A1s of A Company, 2nd Assault Amphibian Battalion, thunder in a line three abreast through the Iraqi desert somewhere in the region of Fallujah. The vehicles belong to a patrol tasked with conducting IED sweeps. Note the coils of razor wire mounted on the hull's bow, useful for quickly setting up surprise traffic checkpoints or roadblocks. (Yves Debay)

Mounted Machine Gun and a 40mm Mk.19 Mod.3 automatic grenade launcher. Today in Iraq, both weapons are a nightmare for any insurgents who find themselves at the receiving end of such lethal firepower. In 1985 the USMC changed the designation of the LVTP7A1 to AAVP7A1. In 1986 the USMC also began increasing the armor protection of the AAVP7A1. As an interim solution, from 1991 onwards, the AAVP7A1 was fitted with the Enhanced Appliqué Armor Kit manufactured by RAFAEL Armament Development Authority in Israel. In 1997 another SLEP was initiated called Reliability, Availability and Maintainability/Rebuild to Standard (RAM/RS) Program. It was aimed at ensuring the AAVP7A1 remained serviceable until the arrival of its replacement, the Advanced Amphibious Assault Vehicle (AAAV) later this decade. The last AAVP7A1 models will be upgraded under the RAM/RS program in 2006.

Amphibious Assault Vehicle

All vehicles of A Company feature the RAM/RS upgrades. Under the RAM/RS Program, the AAVs were fitted with components that are also used in the M2 Bradley Infantry Fighting Vehicle. These include the Cummins VTA-903T 525 8-cylinder water-cooled, turbocharged diesel engine that develops 525hp, and the power train and torsion bars of the Bradley suspension system. The RAM/RS Program involved all vehicles

of the AAVP7A1 family. Today all AAVs, except for the recovery vehicle, are fitted with the Enhanced Appliqué Armor Kit, which adds a weight of some 2,000kg to the vehicle. The bolted-on armor panels increase the vehicle's ballistic protection along the sides of the hull and over the troop compartment. The EAAK provides protection from 7.62mm projectiles, as well as a 95% probability of non-penetration by 12.7mm armor piercing rounds at muzzle velocity. Important for the AAV crews, who in Iraq face deadly IEDs on a daily basis, is the armor's 99% probability of non-penetration from 155mm HE fragments detonating at a range of 15m or higher above the vehicle. As well, the armor substantially decreases the effectiveness of shape-charge weapons by reducing the fragmentation debris cone from 110° to 35°. The EAAK is mounted on the basic hull of the vehicle, which is made of all-welded aluminum and therefore provides only limited protection against small-arms fire and artillery shrapnel in its unmodified state.

Inside the hull, the driver's compartment is situated at the front left. The driver can see his surroundings through seven vision blocks. For night operations a night vision periscope is provided that can be mounted in the one-piece hatch above the driver's compartment. The commander's cupola is placed behind the driver's hatch and an extended periscope allows the

The commander of the AAVP7A1 is seated behind the driver on the left hand side of the vehicle. The vehicle's turret is operated by the third crewmember, the gunner. All three crewmen can be seen in these photos. In the rear compartment of the AAVP7A1 is space for up to twenty-one Marines with all their combat equipment. (Yves Debay)

commander to see over this hatch. Another seven vision blocks provide all-round observation for the commander. In the front right of the vehicle is situated the engine compartment. The water jets (which are situated on both the right and left at the rear of the AAVP7) are driven via a power takeoff using drive shafts. The water jets are used to propel the vehicle forward and also to steer it when operating in water. The capacity of the water jets amounts to a maximum throughput of nearly 53,000 liters of water per minute.

A special two-way water-protected air intake system provides air for the engine as well as for the troop compartment while operating in water or on land. There is no NBC protection system currently fitted to the AAVP7A1 RAM/RS. To prevent the vehicle from sinking in the case of a leak in the hull, four power-driven bilge pumps exist. On the right behind the engine compartment is the "Up-Gunned Weapon Station". The running gear of the vehicle consists of six dual road wheels, a rear idler and a front drive sprocket on each side. Hydraulic shock absorbers are fitted in addition to the torsion bar suspension on the first and last road wheels.

The troop compartment is sited in the rear of the vehicle, and provides space for up to twenty-one fully equipped Marines. They are seated on three benches, one on each side of the hull, and one running down the center of the troop compartment. The benches are folded away when cargo is carried, and up to 4,540kg of cargo can be stored in the 13.56m³ cargo bay. The troops enter and leave the vehicle by a large power-operated rear ramp. There is a door in the left of this large ramp that can be used when the crew chooses not to lower the whole rear ramp. A vision block on the right side of the ramp provides troops in the troop compartment with a limited view to the rear. When the vehicle is used in the armored personnel carrier role, the troops can open the large three-piece hatch over the troop compartment and operate their weapons over the sides of the vehicle.

Company structure at home and abroad

When not deployed, the 2nd Assault Amphibian Battalion, including A Company, is based at Camp Lejeune, North Carolina. It is one of four units in the United States Marine Corps equipped with "Amtracs", which is what the Amphibious Assault Vehicle is sometimes called. The primary task of the battalion is to spearhead beach assaults. In such a scenario, the unit's AAVs will disembark from ships offshore, then ferry troops and equipment ashore; at the same time they will provide armor protection and fire support for the landing force. Once the landing operation has been deemed a success, the AAVs return to the ships or stay with the Marine infantry units and act as armored personnel carriers.

The battalion is structured into a Headquarters and Service Company, and four Assault Amphibian Vehicle Companies (A, B, C and D). The battalion can field 187 AAVP7A1 RAM/RS Personnel Carriers, fifteen AAVC7A1 RAM/RS Command Post vehicles, and five AAVR7A1 RAM/RS Recovery vehicles. In total the battalion numbers some 1,500 Marines, while each AAV company has a strength of 186 Marines. The four companies each consist of a Headquarters and Service Platoon, a Maintenance Platoon, and four AAV Platoons. Each of the AAV Platoons can field fourteen AAVP7A1 RAM/RS vehicles, while the Maintenance Platoon can field one AAVR7A1 RAM/RS, and the Headquarters and

The AAVP7A1 looks more like a ship on tracks than an armored vehicle! Marines said of the vehicle: "We often benefit from the height of the vehicle. We can see down into places and from angles that a HMMWV crew cannot. In addition, the Iraqis have huge respect for us - the size and noise of the AAVP7A1 RAM/RS scares them and they fear our firepower." (Yves Debay)

AAVs 'steam' through the desert sands of western Iraq, kicking up a wake of clouds of fine dust. The AAV is powered by a Cummins VTA-903T 525 8-cylinder water-cooled, turbocharged diesel engine that develops 525hp. This gives a creditable top speed of 72.5km/h on land. (Yves Debay)

AAVP7A1s of A Company "Gators", 2nd Assault Amphibian Battalion, somewhere outside Fallujah in December 2005. The company can field a total of forty-four monstrous armored amphibious assault vehicles. In Iraq the company supports the infantry units of RCT-8 with its armor-protected transport capability. (Yves Debay)

The message is clear – stop! The bilingual traffic sign has been attached to the front of the vehicle, and would be useful in the traffic checkpoint or roadblock type of missions that AAVs are commonly engaged in. This AAV also has a tire mounted on the bow – it is obviously not a spare tire, but functions in the same way as tires seen on the sides of ships! (Yves Debay)

Service Platoon can field two AAVC7A1 RAM/RS and two AAVP7A1 RAM/RS. Also belonging to the Headquarters and Service Platoon are five HMMWVs, two MTVR 7-ton trucks and two LVS trucks, one in a cargo and one in a fuel tanker configuration.

Mission-oriented structure

For Operation Iraqi Freedom III, A Company of the 2nd Assault Amphibian Battalion restructured in order to form a fourth operational element. From each of the three AAV platoons, three vehicles were taken and assigned to the Headquarters and Service Platoon. Together with the two original AAVP7A1 RAM/RS, an additional AAV platoon was thus formed. As a result, the company features four equally structured operational assets and a reduced Headquarter and Service element. Each of the four new AAV platoons is divided into three sections, two with four AAVs and one with three AAVs.

"In addition to supporting infantry units during operations, our platoons also conduct their own missions if required," explained Lt. Ross, an AAV platoon commander. He added: "In this case we only operate with six AAVs from a platoon; the rest of the crews re-role into infantrymen. In a platoon that makes some twenty-four dismounts as an AAV crew can be between three and five Marines. In this way we conducted some quite effective sweep operations during which some weapon caches were discovered. We also did cordon and knock operations, Main Supply Route security operations, IED sweeps and vehicle checkpoints."

While most units belonging to RCT-8 have been assigned their own Areas of Operation, A Company, 2nd Amphibious Assault Battalion, has not. The unit is subordinated directly to RCT-8 Headquarters and is assigned wherever its capabilities are needed. This might even be in support of other units of the II Marine Expeditionary Force outside RCT-8's AOO. On average, the company in Iraq is called upon three times a month to support a large-scale operation. For most of the Marines of A Company, it is the second time that they have deployed to Iraq. The company participated in Operation Iraqi Freedom I, the initial invasion of Iraq. For Operation Iraqi Freedom III, the company received personnel support from the 2nd Assault Amphibian Battalion's C and D Companies. Of these troops, some 40-50% had also deployed to Operation Iraqi Freedom II. This means that the "Gators" (the radio net nickname for Marines of A Company) have a lot of experience with which to fight Iraqi insurgents. Other recent operations and deployments saw A Company operating without their AAVs in Afghanistan in 2004, taking part in "Exercise Bright Star 2001", as well as deploying to Kosovo in 1999.

Grueling experience

On 12 October 2005, during "Operation Liberty Express", the 1st Platoon, A Company, 2nd Amphibious Assault Battalion, was conducting a twelve-hour patrol. Just twenty minutes before the end of the patrol it happened. The vehicles of the platoon were moving in a column down a road following some civilian cars, which proceeded to move aside to let the Amtracs pass. As the second of the three AAVP7A1 RAM/RS vehicles was passing the cars, one car pulled out and darted in the direction of this AAV. Just 10-15 feet in front of it, the suicide attacker inside the car blew himself up. "The AAV was enveloped in a ball of flames - I thought it was gone," said Gunnery Sergeant Green, who was traveling in the third AAV. He added: "But except for minor damage, it was fine! We could even drive it home on its own tracks." However, the three crewmembers that had been driving with their hatches open were seriously wounded and they needed to be evacuated. The car that had been used as a SVBIED had totally

Another frontal view of an AAV. This vehicle has had its trim vane removed from the front of the hull, it not really being necessary in the desert sands of Al Anbar province. The vehicle's maximum speed in water is 13.2km/h, and with its 647-liter fuel tank, it can operate in the water for up to seven hours. (Yves Debay)

The AAVP7A1 RAM/RS borrowed a number of components from the successful Bradley operated by the U.S. Army. Some of these items include the Cummins VTA-903T 525 8-cylinder water-cooled, turbocharged diesel engine, and the power train and torsion bars of the Bradley suspension system. The vehicle rides on six road wheels, plus there are return rollers and a front-mounted drive sprocket. (Yves Debay)

disintegrated. The vehicle that was attacked was "2A111", which stands for 2Bn, A Company, 1st Platoon, 11th vehicle.

But not every attack on Amtracs in Iraq has ended so luckily for the crew. On 3 August 2005, an Amtrac of the 4th Assault Amphibian Battalion hit a roadside IED made of three stacked antitank mines just outside Haditha. The blast nearly ripped the vehicle to pieces. The hull in the area of the rear troop compartment was gashed open like a 'V', while the vehicle was tossed into the air and landed on its roof some 10m away. The fourteen Marines of Regimental Combat Team 2 traveling inside the vehicle were killed instantly. The incident caused serious discussions about the AAVP7A1 RAM/RS, and whether or not the vehicle was suitable for operations in Iraq. Concerning this incident, it must be stated that even the most heavily armored main battle tank would have been destroyed by such a powerful blast. We would like to finish our article with some words from Gunnery Sergeant Green: "It might not be the best of vehicles in the theater, but during operations we often benefit from the height of the vehicle. We can see down into places and from angles that a HMMWV crew cannot. In addition, the Iraqis have huge respect for us - the size and noise of the AAVP7A1 RAM/RS scares them and they fear the firepower of the 12.7mm machine gun and 40mm automatic grenade launcher in the turret." The motto of the Marines of A Company, 2nd Amphibious Assault Battalion, clearly indicates their feelings for the AAV7A1 RAM/RS: "YAT YAS – You Ain't Tracks, You Ain't Shit."

This is another AAV photographed at the motor pool. The vehicle is a dust-streaked green color, but the EAAK is sand-colored. There is a number "30" stenciled on the front of the EAAK. Note the soft toy mascot mounted on the front of the vehicle. (Yves Debay)

Another form of traffic sign appears on this AAV. The weapons have been removed from the turret since the vehicle has been left parked at the battalion's motor pool. A complete assault amphibian battalion can field 187 AAVP7A1 RAM/RS Personnel Carriers, fifteen AAVC7A1 RAM/RS Command Post vehicles, and five AAVR7A1 RAM/RS Recovery vehicles. (Yves Debay)

Another family member is the AAVC7A1. Like the AAVP7A1, the vehicle is fitted with the RAFAEL EAAK. In evidence are the antennas for the vehicle's extensive radio network. The Command Post's communications suite comprises one UHF, one HF, and six VHF radios. In the rear compartment of the vehicle are workstations for five staff officers and five radio operators. These ten passengers can be transported in addition to the three-man crew. (Yves Debay)

Side views of two different AAVP7A1s of A Company, seen during a mission in Iraq. Visible is the EAAK. The armor adds some 2,000kg to the vehicle's weight. The EAAK provides protection from 7.62mm and smaller projectiles, as well as a 95% probability of non-penetration by 12.7mm armor piercing rounds at muzzle velocity. It also gives a 99% probability of non-penetration from 155mm HE fragments detonating at a distance of 15m overhead. Also, the armor decreases the effectiveness of shape-charge weapons by reducing the fragmentation debris cone from 110° to 35°. (Yves Debay)

A frontal view of a "tuna boat", as this Marine vehicle is sometimes referred to. Of note is the wire cutter fitted to the bow to protect crewmembers from obstacles such as low-hanging wires. These devices have only been fitted to some vehicles. (Yves Debay)

Technical Data for AAVP7A1 RAM/RS

Crew:	3 (driver, commander and gunner) + up to 21 fully equipped Marines in the troop compartment
Length:	8.16m
Width:	3.27m without EAAK
Height:	3.31m
Unloaded weight:	21,027kg
Combat weight:	25,761kg (with troops loaded but without EAAK)
Load capacity:	21 combat equipped troops (each approximately 129kg) or 4,540kg of cargo
Maximum speed on land:	72.5km/h
Maximum speed in water:	13.2km/h
Cruising range:	322km at a speed of 40km/h or 7 hours in water
Fuel tank capacity:	647 liters
Ground clearance:	406mm
Vertical obstacle:	910mm
Trench crossing:	2.44m
Engine:	Cummins VTA-903T 525 8-cylinder water-cooled, turbocharged diesel engine developing 525hp
Transmission:	NAVSEA HS-525 with four forward and two reverse gears
Armament:	1x 12.7mm M48 Turret-Mounted Machine Gun
	1x 40mm Mk.19 Mod.3 automatic grenade launcher
	A pair of 4 x 66mm M257 Smoke Grenade Launchers
Gun elevation/depression:	+60° to -15°
Gradient:	60%
Side slope:	40%
Cargo compartment:	Length: 4.12m; Width: 1.83m; Height: 1.67m; Volume: 13.56m³
Variants:	AAVC7A1 RAM/RS Command and Control Vehicle
	AAVR7A1 RAM/RS Recovery Vehicle.
Manufacturer:	United Defense LP, belonging to BAE Systems

The HQ of "Gator" Battalion, as the 2nd Assault Amphibian Battalion is known. This photo shows the battalion's logo with its appropriate alligator character. (Yves Debay)

This M60A1 AVLB was spotted at the 2nd Tank Battalion motor pool. The bridge-layer can launch and retrieve a 60-foot scissors bridge, and vehicles as heavy as 60 tons can cross over it. These engineering vehicles are now showing their age, and in the initial invasion of Iraq, many just could not keep pace with the rapid advance. (Yves Debay)

The AAVR7A1 RAM/RS is the only version of the Amtrac not fitted with the EAAK. The vehicle is fitted with a hydraulic telescoping-boom crane with a 2,700kg lift capacity, and a 13,500kg capacity recovery winch. The vehicle has a five-man crew, and welding and cutting tools are part of the specialized on-board equipment. (Yves Debay)

Not a sight that an insurgent would relish! AAVP7A1s roar towards the camera, with clouds of sand and dust billowing behind them. The automatic weapons in the UGWS include the 0.50-cal M48 Turret-Mounted Machine Gun and 40mm Mk.19 Mod.3 automatic grenade launcher. (Yves Debay)

The rear compartment of the AAVC7A1 RAM/RS is equipped with one HF, one UHF, and six VHF radios. There are workstations available for five staff officers and five radio operators. The dimensions of the cavernous troop compartment of the AAV are 4.12m long, 1.83m wide and 1.67m high. This gives an internal volume of 13.56m³. (Yves Debay)

These close-ups from different angles show to advantage the "Up-Gunned Weapon Station" of an AAVP7A1 RAM/RS. A pair of 4 x 66mm M257 smoke grenade launchers is located at the rear of the one-man turret. (Yves Debay)

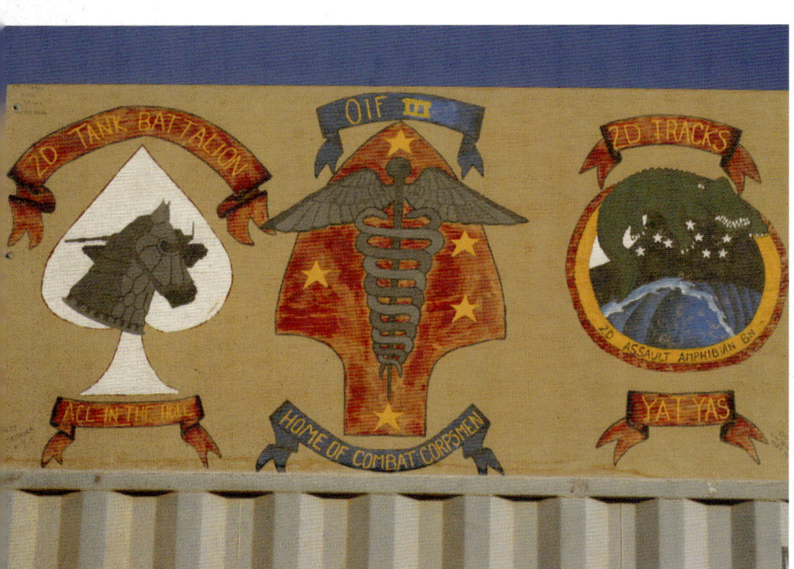

A shot of the large sign displaying the resident units located at A Company, 2nd AABN's camp near Fallujah. They include 2nd Tank Battalion and a Corpsmen medical unit. (Yves Debay)

"European Challenge" for the Bundeswehr

Carl Schulze

A pair of Fennek 4x4 wheeled reconnaissance vehicles of Panzeraufklärungsbataillon 5 use their mast-mounted sensor pods to gather information on local militia that are preparing an attack on an IFAB peace force checkpoint. The sensor equipment includes a laser rangefinder, thermal sight and daytime TV camera. The communications equipment of the Fennek consists of a SEM 80/90 VHF radio and an HRM 7400 HF radio.

During the Cold War Germany played host to large annual field training exercises involving tens of thousands of troops from several NATO nations. Divisions and even corps trained with 100% of their strength in the open countryside. Attacks involving hundreds of tanks and armored personnel carriers were a common sight in West Germany. U.S. Forces trained in Return of Forces to Germany (REFORGER) exercises with the deployment of whole corps in order to counter the perceived Warsaw Pact threat. But now times have changed. Some fifteen years after the crumbling of the Warsaw Pact and the fall of the Iron Curtain, these large-scale exercises have vanished. Seldom do troops train in the open countryside, and if they do so, the exercise is usually brigade sized only. Exercises are conducted more and more in special training areas or are of the computer assisted command post type. In addition, the scenarios have changed from fighting a high intensity conflict against an overwhelming enemy force, to establishing peace in a crisis hit region while facing irregular forces. One of the largest exercises to be held in Germany for years was Exercise "European Challenge 2005".

Joint combined training

Between 18th and 29th April 2005, German armed forces conducted the joint combined Exercise European Challenge 2005. This exercise involved 4,500 troops and was set up to simulate a crisis management scenario. The majority of the exercise was conducted as a command post exercise (CPX). In the CPX the Force HQ, which Germany places under EU command, was the main player. The Force HQ is based on the HQ of the II German/U.S. Corps.

Subordinated to the Force HQ were an Air Component Command (ACC) provided by the German Air Forces Command (Luftwaffenführungskommando), a Maritime Component Command (MCC) provided by the German Naval Forces Command (Flottenkommando), a Psychological Operations Command provided by the German Centre for Psychological Operations (Zentrum Operative Information), a Special Operations Component Command (SOCC) provided by the Division Special Operations (Division Spezielle Operationen), and an Operational Command

Paratroopers of Luftlandebrigade 26 have just landed in order to prepare for a non-combatant evacuation. The soldier in front is armed with a 5.56mm G36 assault rifle and a Panzerfaust 3 anti-tank weapon. Note the new standard issue combat knife sticking out from under his combat vest.

This paratrooper is equipped with the latest German infantry equipment called "Infanterist der Zukunft". He is armed with a 5.56mm G36 assault rifle, and mounted on his helmet can be seen a pair of "Lucie" night vision goggles. Note that the G36 is fitted with a bipod. This soldier belongs to Fallschirmjägerbataillon 263.

Close-up of the 40mm Heckler & Koch automatic grenade launcher mounted on the Mercedes 250 GD of Fallschirmjägerbataillon 263. The weapon is fitted with a reflex sight. Operating with a blowback mechanism, the weapon has a theoretical rate of fire of 350 rounds per minute.

The 750 soldiers of the evacuation operation force were mainly troops from Fallschirmjägerbataillon 263 (airborne battalion). Here a Mercedes 230G of the battalion can be seen. The vehicle is armed with the newly introduced Heckler & Koch 40mm automatic grenade launcher. Together with a tripod and a box with 32 rounds, the weapon has a weight of 75.5kg.

(Einsatzführungskommando). Under the LCC (Land Component Command) the HQ element of the German Airmobile Division (Division Luftbewegliche Operationen - DLO) acted as a division level HQ. The DLO was responsible for commanding the brigade level HQs involved in Exercise European Challenge 2005. This included the HQ of Jägerbrigade 37, an Austrian mechanised brigade HQ, a Czech airborne brigade HQ and the HQ of the German Luftbewegliche Brigade (airmobile brigade). Under the SOCC Luftlandebrigade 26 (airborne brigade) acted as brigade level HQ. Within the HQs taking part in the exercise, personnel from fifteen EU and NATO nations were integrated.

The CPX part was conducted at the Combat Simulation Centre (Gefechtsimulationszentrum) of the German Army in Wildflecken, southern Germany. During the CPX the HQs had to deal with a complex crisis management scenario revolving around several different ethnic groups, ethnic cleansing, terrorist activities, disputed territory and a fragile peace between two former warring states, Beachland and Amberland. The aim of the exercise was to train for the coordination of land, air and naval forces utilising all of the modern equipment available to the Bundeswehr. An additional aim was to train in joint combined operations in a "Three Block War" setting, ranging from the provision of humanitarian aid, the separation of warring factions, and demilitarisation, right up to high intensity war fighting. Overall command and control for Exercise European Challenge 2005 was maintained by the German Ground Forces Command (Heeresführungskommando) and its commander, General Bürgener. European Challenge 2005 was the first large-scale exercise in which the German Armed Forces, consisting of Army, Navy, Air Force, Medical Service and Support Service, had trained together in an EU-led crisis reaction scenario.

A CH-53G(S) of the German Heeresflieger (Army Air Corps) returns with evacuated civilians to the FOB of Luftlandebrigade 26 during the non-combatant evacuation operation. The CH-53G(S) is the latest model of the medium transport helicopter to enter German service. The helicopter features external fuel tanks that extend the aircraft's operational range from 1 hour 40 minutes to 6 hours 30 minutes. The CH-53G(S) has a maximum takeoff weight of 19 tons and can transport up to thirty-six soldiers or 5.5 tons of cargo.

Capability demonstrations

Embedded within Exercise European Challenge 2005 were two capability demonstrations in which the Bundeswehr showed its latest equipment and demonstrated its joint combined capability. On 19th April 2005, Luftlandebrigade 26, together with navy and air force assets, demonstrated the evacuation of civilians out of a crisis area in Flensburg. The demonstration formed part of the brigade's Field Training Exercise (FTX) that was conducted within the framework of Exercise European Challenge. On 20th April 2005, Jägerbrigade 37 practised elements of a crisis

A dog handler and his dog of Fallschirmjägerbataillon 263 await orders to search a couple of bags for explosives and hidden weapons. The dog handler platoons were recently introduced into German airborne battalion structures. The reasons for this were the new security threats and new types of missions faced by the paratroopers. While in the past they were seen as an intervention force that fought behind enemy lines, today the paratroopers' main mission is to operate as an early entry or crisis response force in a peace support operation scenario.

management operation in a live fire exercise. The two-hour demonstration was held at the Bergen Hohne ranges and involved standard peace support operation components. These included such elements as maintaining checkpoints, suppressing a civil uprising and providing humanitarian aid, as well as the use of an all-arms battle group supported by combat aircraft designed to push back a hostile force attacking the deployed peace support forces.

Non-combatant evacuation operation

The FTX of Luftlandebrigade 26 was conducted in northern Germany and involved some 1,350 troops from brigade and support units which combined to form an Evacuation Operation Force. The exercise took place from 4 – 22 April 2004. The exercise involved all the parts of a Non-

This paratrooper of Fallschirmjägerbataillon 263 belongs to the security perimeter drawn around a collection point for civilians soon to be evacuated back to Germany. He is armed with a 5.56mm G36 assault rifle. Note the laser light module fitted to the weapon. The LLM features a visible as well as a non-visible target pointer and a light source.

A dog searches baggage belonging to non-combatants for explosives. This is a procedure necessary to prevent any suicide attacks aimed at civilians who will soon be removed from the crisis zone. The dog handler belongs to Fallschirmjägerbataillon 263.

Combatant Evacuation Operation (NEO), from gathering the forces, moving the forces to a Forward Mounting Base (FMB) at Jever, deploying them and establishing a Forward Operating Base (FOB) in Tarp/Eggebeck, executing the evacuation and then conducting the redeployment of the forces. The NEO was a joint exercise involving German Air Force assets as well as ships of the German Navy. The NEO force was formed from units of Luftlandebrigade 26. A prime asset was Fallschirmjägerbataillon 263, which was reinforced by elements of Luftlandeunterstützungsbataillon 262 and other non-brigade assets. These assets included elements of the Military Police, an Electronic Warfare element, a reconnaissance element, an army aviation element with six CH-53G medium transport helicopters and six Bell UH-1D light transport helicopters, NBC defense elements, and elements of a psychological operations unit.

The exercise began on 4th April with the evacuation force being placed on standby at short notice. On 8th April the force deployed from their home bases to a Staging Area at the training area in Darden, which was situated close to an Air Port of Embarkation. In Darden the NEO force conducted intensive mission orientated pre-operation training for the next three days. This training was conducted and prepared by Luftlandeunterstützungsbataillon 262. Between 11th and 15th April the NEO force deployed via a strategic air move to Jever. The airport and the surrounding country acted as parts of a friendly country and were used as a Forward Mounting Base.

17th April saw the NEO force deploying their reconnaissance assets into crisis-shaken Beachland. At the same time, embassy personnel gathered German citizens and moved them to three assembly points. 19th April witnessed the tactical deployment of the 750 strong NEO force into the Forward Operation Base. After the FOB was secured, the troops immediately began preparations for the evacuation. At the same time

Here a Wiesel 1 equipped with a 20mm Mk 20-2 machine cannon can be seen guarding the FOB of Luftlandebrigade 26. The FOW was established at the German Naval Air Corps station Tarp/Eggebeck. The Wiesel 1 was deployed to a sentry position in order to protect the runway against enemy attacks.

German Navy ships reached their positions in front of the coast of Beachland in order to take on board evacuated citizens. The plan was to have all Germans evacuated by 23rd April as this day marked the anniversary of the death of an Amberland enclave militia leader. It was suspected that the anniversary would lead to major unrest. Therefore, this meant German paratroopers were busy extracting non-combatants through the use of helicopter and wheeled transport on October 21st and 22nd. Once taken to the secure FOB, the civilians were then evacuated by ship or aircraft. The afternoon of 22nd April also brought the extraction of the non-combatant evacuation force to the FMB. At the FMB the end of the exercise was announced and it was declared as a mission well done. In total the German paratroopers evacuated 95 civilians

Jägerbrigade 37 "Freistaat Sachsen"

Jägerbrigade 37 "Freistaat Sachsen" was formed on 1st April 1991. During the brigade's official inauguration ceremony, the brigade was given the name "Freistaat Sachsen" (Independent State Saxon). Originally raised as a home defense brigade in Dresden, the formation was soon reconstituted into a mechanised infantry brigade and moved its HQ to Frankenberg in 1995. Under the "Army for new missions" structure, the brigade was again reformed, and beginning in 1996 became a light infantry brigade consisting of an airborne infantry battalion (Fallschirmjägerbataillon 373), a light infantry battalion (Jägerbataillon 371) and a mountain infantry battalion (Gebirgsjägerbataillon 571). The brigade was assigned to the crisis reaction forces and became operational in 1998. The brigade was also responsible for keeping an infantry battle group ready for deployment with the Allied Command Europe Mobile Force (Land), AMF(L). The AMF(L) unit was provided by the mountain infantry battalion of the brigade.

A detail picture of the sensor pack of a Wiesel 1 airmobile reconnaissance vehicle. The elevation mast-mounted sensor pack consists of a thermal camera and a laser rangefinder.

A CH-53G(S) has just landed at the FOB of the non-combatant evacuation operation force in order to deliver wounded troops to a forward deployed medical facility. Usually situated in the FOB, the medics and doctors have all the necessary equipment to stabilise even badly wounded personnel. The CH-53G(S) has a crew of four, comprising of a pilot, a co-pilot and two loadmasters. With external fuel tanks the aircraft has a range of 650km.

In the following years the brigade deployed twelve times with battalion-sized units to Bosnia, Kosovo and Afghanistan as part of SFOR, KFOR and ISAF respectively. The year 2004 saw soldiers of the brigade on duty in Kabul and Kunduz. In addition, soldiers of the brigade were regularly involved in disaster relief operations in Germany, for example the flood crises in 1997 and 2002. Currently the German Army is again changing its structure in order to be better prepared for future operations. In this process

For casualty evacuation the medical elements of airborne infantry battalions of the Bundeswehr have received an ambulance version of the Wiesel 2 airmobile vehicle. The vehicle can be transported by a CH-53G medium transport helicopter, either as an internal or under-slung load. One wounded soldier on a stretcher can be carried inside the Wiesel 2 medical vehicle. The pictured vehicle belongs to the 1st Company of Fallschirmjägerbataillon 313.

the brigade lost its airborne infantry battalion, but gained command of a tank battalion (Panzerbataillon 393), and a mechanised infantry battalion (Panzergrenadierbataillon 391). In the future the brigade will disband its mountain infantry battalion and reform its light infantry battalion into a mechanised infantry battalion. By 2010 the brigade will have completed its transformation into a mechanised infantry brigade.

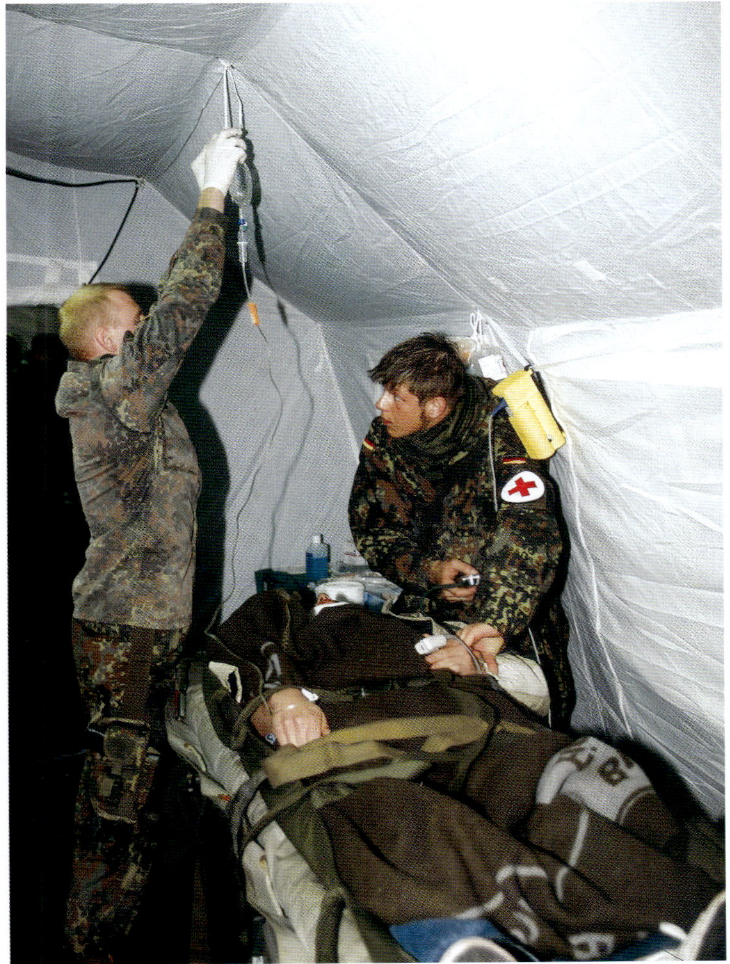

One element of the 750 strong non-combatant evacuation force was an airmobile first aid station. The medical facility has the equipment to stabilise even critically injured patients so that they can be evacuated. Here medical personal can be seen looking after a civilian who was injured when his car was attacked by insurgents.

A rear view of a Wiesel 2 medical vehicle with its doors open. Inside the rear compartment of the vehicle one wounded soldier on a stretcher can be carried. The pictured vehicle belongs to the 1st Company of Fallschirmjägerbataillon 263.

A Transall C-160 has landed in the FARP to lift out the first batch of non-combatants to the Forward Mounting Base. A military policeman checks the passenger list of the aircraft. The civilians have just been brought to the runway by Unimog 2-ton trucks of Fallschirmjägerbataillon 263.

From the back of a Unimog 2-ton truck, a paratrooper of Fallschirmjägerbataillon 263 scans his surroundings for any threat. He is armed with a 5.56mm G36 assault rifle. Note that the weapon's butt has been folded.

Dog handlers of Fallschirmjägerbataillon 263 have deployed and are now kept in reserve as local workers protest against the evacuation of German citizens. Today each German airborne battalion has a dog handler platoon in its ORBAT. The platoon can field guard dogs as well as dogs that can search for weapons, explosives, drugs and so on.

This Mercedes Benz MD 250 4x4 cross-country vehicle belongs to Fallschirmjägerbataillon 263. The German paratroopers have mounted a Milan guided anti-tank weapon system on the roll cage of the vehicle. The Milan mount is a workshop conversion designed and produced at the unit level.

During the non-combatant evacuation operation, rescued civilians were moved to a secure Baltic Sea port and evacuated by German Navy ships. Here a paratrooper of Fallschirmjägerbataillon 263 can be seen checking the number of civilians just brought in by his comrades. The soldier is armed with the 5.56mm G36 assault rifle and wears the new German airborne forces helmet.

Civilians rescued by the German paratroopers of Fallschirmjägerbataillon 263 board a ship of the German Navy in a Baltic Sea port. Non-combatant evacuation operations can be conducted using air, naval and ground forces assets. Evacuation is most likely to be done by air or sea, however.

One of the main aims of Exercise European Challenge 2005 was to test joint combined structures. During the demonstrations therefore, army, navy and air force assets worked closely together. Here the German Type 122 Frigate Niedersachsen F208 can be seen patrolling the coast while paratroopers conduct an evacuation operation. The ship's armament includes eight Harpoon SSMs and an eight-cell Sea Sparrow SAM launcher.

The capability demonstration in Eckerförde on 19th April also involved members of the German combat diver company. Though usually operating under the cover of darkness, for this demonstration the naval Special Forces were brought out into the daylight! The divers are armed with standard 5.56mm G36 assault rifles.

A Wiesel 1 airborne weapon carrier armed with a 20mm Fk 20-2 cannon of Fallschirmjägerbataillon 263 moves out of its recently occupied firing position. This photo was taken during the non-combatant evacuation operation performed in Eckernförde on 19th April 2005. The Wiesel 1 machine cannon version has a crew of two, composed of a driver and a commander. The 20mm cannon has a combat range of 1,200m. The depicted vehicle is one of the upgraded models featuring a thermal imaging sight.

In the future each Luftlandebrigade of the Bundeswehr will have a reconnaissance company under its command. The company will be equipped with the reconnaissance version of the Wiesel 1 airmobile vehicle. Here one of the first vehicles ever delivered can be seen. The vehicle belongs to Luftlandeaufklärungskompanie 310. In total, ten Wiesel 1 TOW airmobile weapon carriers were converted into Wiesel 1 reconnaissance vehicles. The major change was the removal of the TOW system and the addition of a raised armored roof that incorporates a short mast and the sensor package.

A Forward Air Control party of Luftlandebrigade 31 guides a C-160 Transall transport aircraft to a humanitarian goods drop zone. During the live fire exercise at Bergen Hohne, humanitarian goods were dropped to aid the civilian population that was suffering from the impact of an ethnically related conflict. The weapon in the foreground is a 5.56mm G36k A1 assault rifle.

A pair of Wiesel 1 airmobile reconnaissance vehicles has just arrived at a designated helicopter landing site in order to be extracted following a reconnaissance mission. Two of these vehicles can be loaded into one CH-53G medium transport helicopter. The vehicles belong to Luftlandeaufklärungskompanie 310. Arming the Wiesel 1 reconnaissance vehicle is a 7.62mm MG3 machine gun. The vehicle has a three-man crew and is powered by a VW 5-cylinder diesel engine.

Led by a TPz 1A4 Fuchs 6x6 wheeled armored personnel carrier and an ATF Dingo 1 protected carrier vehicle, a humanitarian aid convoy of Jägerbrigade 37 can be seen making its way to a Beachland village desperately in need of medicine and food.

The ATF2 Dingo 1 protected carrier vehicle was introduced into the Germany Army in 2000. Today the mine-protected vehicle is used in Kososvo and Afghanistan by German forces for patrolling and convoy escort duties. The vehicle is fitted with a roof-mounted weapon station that can be operated by the crew from under armor. The weapon station can be fitted with either a 7.62mm MG3 machine gun or a 40mm automatic grenade launcher.

This TPz 1A4 Fuchs 6x6 wheeled armored personnel carrier is used by an armored reconnaissance battalion and is fitted with the battlefield surveillance radar RASIT. The radar can detect and identify movements on the battlefield, a capability that is also very useful during peace support operations. The radar can be raised up to 1.8m above the hull roof. This vehicle belongs to Panzeraufklärungsbataillon 5.

This TPz 1A4 Fuchs 6x6 wheeled armored personnel carrier is used by Panzerpionierkompanie 390 to guard a static vehicle checkpoint in the demilitarised zone that separates Amberland and Beachland. The amphibious vehicle is powered by a Mercedes Benz 8-cylinder diesel engine that develops 320hp. The vehicle can reach a maximum road speed of 105km/h, and when used in an amphibious environment it can reach a top speed of 10.5km/h. The Fuchs will be replaced by the 8x8 Boxer in the future.

Leopard 2A6 tanks advance to contact in order to counter a threat against peace support forces by an enemy armored formation. The Leopard 2A6 is the latest version of the Leopard 2 family and is armed with a 120mm L55 smoothbore gun. Powering the Leopard 2A6 is an MTU Mb 873 12-cylinder diesel engine, which allows the vehicle to reach a top speed of 72km/h.

Later the Bundeswehr will only maintain six tank battalions. Most of these battalions will be equipped with the Leopard 2A6. Here a platoon of Leopard 2A6s of Panzerbataillon 383 can be seen preparing for a counterattack. The Leopard 2A6 has a combat weight of 60 tons and is armed with the new 120mm L55 smoothbore gun. Inside the vehicle 42 rounds for the gun can be stored.

A Panzerhaubitze 2000 of Panzerartilleriebataillon 115 provides fire support during the counterattack of armored units under the command of Jägerbrigade 37. The effective range of the Panzerhaubitze 2000 with standard 155mm artillery ammunition is 30km.

The Panzerhaubitze 2000 is equipped with a 155mm gun with a 52-calibre barrel. In emergencies the gun can be used for direct firing at a range of up to 2,000m. The PzH 2000 has a combat weight of 55 tons and is powered by an MTU 881 diesel engine. Sixty 155mm rounds for the gun can be carried inside the vehicle.

Here the fighting compartment of the 56-ton Panzerhaubitze 2000 self-propelled howitzer can be seen just before a fire mission is executed. With its automatic loading system the Panzerhaubitze 2000 is able to fire a burst of three 155mm rounds in less than ten seconds. Intensive fire can also be provided by firing eight rounds per minute. Thus in three minutes up to 20 rounds can be fired.

The commander of a PzH 2000 is checking a fire command for his 155mm self-propelled howitzer on his commander's display.

In recent years the German Army has discarded their ageing fleet of M113 artillery observation vehicles. As a replacement the FOOs received modified Marder 1A3 infantry fighting vehicles. The pictured FOO belongs to Panzerartilleriebataillon 115.

The Marder 1A3 IFV is the vehicle used by German mechanised infantry, also known as Panzergrenadiere. The pictured vehicles belong to Panzergrenadierbataillon 122. Armed with a 20mm cannon, a 7.62mm machine gun and a Milan guided anti-tank weapon system, the Marder 1A3 is the latest version of the vehicle originally introduced 30 years ago.

A Marder 1A3 armored infantry fighting vehicle can be seen covering a humanitarian aid drop zone from its secure position. During the live fire exercise at the Bergen Hohne ranges, supplies were dropped out of a C-160 Transall transport aircraft from a low altitude. This Marder 1A3 AIFV belongs to the 2nd Company of Panzergrenadierbataillon 122.

Close air defense for the troops of Jägerbrigade 37 was provided by elements of Leichte Flugabwehrraketenbatterie 100 with the Ozelot light air defense system LeFlaSys. The Ozelot fires Stinger missiles and is based on the Wiesel 2 airmobile vehicle chassis. The launcher unit of the vehicle can be rotated a full 360° and has an elevation from −10° to +70°. The launcher unit consists of a sensor and electronic element, a thermal imaging camera, a laser rangefinder, a TV camera, a line-of-sight stabilisation unit, an auto-tracking unit and four launchers.

Also based on the Wiesel 2 airmobile vehicle chassis is the radar, surveillance and fire control vehicle (AFF Wiesel) of the LeFla Sys. Here the vehicle can be seen with its HARD Radar erected. The radar is able to identify targets up to a range of 20km away and up to a height of 5,000m. It can track up to 20 targets simultaneously and can distinguish between fixed wing aircraft and helicopters. The vehicle belongs to Leichte Flugabwehrraketenbatterie 100 and is linked with the Ozelot launcher vehicle by data communications.

The Luchs 8x8 wheeled armored reconnaissance vehicle has been in use for 30 years among Bundeswehr reconnaissance units. Soon this vehicle armed with a 20mm machine cannon will be replaced by the new Fennek wheeled reconnaissance vehicle. The pictured vehicle belongs to Panzeraufklärungskompanie 370.

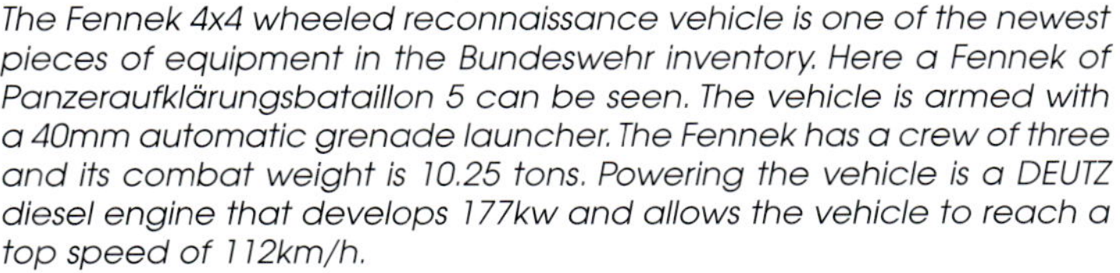

The Fennek 4x4 wheeled reconnaissance vehicle is one of the newest pieces of equipment in the Bundeswehr inventory. Here a Fennek of Panzeraufklärungsbataillon 5 can be seen. The vehicle is armed with a 40mm automatic grenade launcher. The Fennek has a crew of three and its combat weight is 10.25 tons. Powering the vehicle is a DEUTZ diesel engine that develops 177kw and allows the vehicle to reach a top speed of 112km/h.

Light infantrymen of Jägerbataillon 371 have taken up a position in order to defend their checkpoint against an enemy attack. The first soldier is armed with a 5.56mm MG4 light machine gun. On his weapon a Hunt IR thermal sight is mounted. The MG4 will replace the MG3 at a rate of two to one at the section level.

Belonging to Jägerbataillon 371, this light infantryman is equipped with the very latest German infantry combat equipment. He is armed with a 5.56mm G36 assault rifle and wears fragmentation protection goggles, new body armor and a load bearing vest. The new equipment was issued under the program name of "Infanterist der Zukunft."

A Milan anti-tank guided weapon team of Jägerbataillon 371 loads its weapon system in order to counter an enemy tank attack. Milan is the prime medium range anti-armor weapon used by German infantry. The weapon has a combat range of 2,000m and is guided by copper guidance wires. The Milan firing post has a weight of 16.9kg while a missile in its container has a weight of some 13kg.

In a defensive dugout position, light infantrymen of Jägerbataillon 371 use a 40mm automatic grenade launcher to keep an enemy attack at bay. This weapon can fire high explosive, smoke, illumination and armour piercing rounds up to a range of 2,000m.

An infantryman of Jägerbataillon 371 is in position at an IFAB checkpoint. He is armed with the 5.56mm G36 assault rifle, and it is fitted with a Laser Light Module (LLM) and a 100-round drum magazine. His personal equipment includes fragmentation protection goggles, new body armour and a load bearing vest. The equipment was issued under the program name "Infanterist der Zukunft".

A German Army Air Corps Bell UH-1D light transport helicopter delivers Austrian infantry in order to reinforce a checkpoint of Jägerbrigade 37 during the live fire exercise in Bergen. The ageing Bell UH-1D is soon to be replaced by the NH-90 utility helicopter.

The Luna UAV has just entered service with the Bundeswehr and is used to gather virtually real-time information at the brigade level. Here a Luna UAV of Artillerieaufklärungsbataillon 121 (artillery observation battalion) is ready to takeoff in order to support Jägerbrigade 37. The Luna has an effective range of 40km and flight duration time of up to two hours. As seen here Luna is launched by the use of a rubber band catapult. Once airborne the system is powered by a two-cylinder two-stroke petrol engine via a three-bladed propeller.

Equipped with anti-riot gear, soldiers of the light infantry Jägerbataillon 371 prepare to push back a violent mob using batons and shields. Note the video documentation team of the German Military Police positioned on the Fuchs wheeled APC in the background. This team belongs to the 4th Company, Feldjägerbataillon 351 and records the riots on film. This recorded material can later be used to identify ringleaders as well as provide evidence in front of a court of law.

The "Tiger" multi-role combat helicopter will replace the BO105P in German Army service. The first Tigers are currently on delivery to the Heeresflieger (Army Air Corps). The Tiger has a crew of two, consisting of pilot and gunner. The helicopter can be armed with a large variety of weapons including Stinger SAMs, unguided rockets, 12.7mm machine guns and different anti-tank guided missiles. A mast-mounted sensor pack is situated on the rotor, with the pack including a day/night thermal camera system and a laser rangefinder. (Clemens Niesner)

Inside the ground control station of a Luna UAV, the aircraft operator can be seen exercising control over the aircraft during a reconnaissance flight. The control station is based on a Mercedes Benz 290 GD 4x4 cross-country vehicle.

In order to make the conduct of the live exercise at Bergen Hohne more realistic, the Jägerbrigade 37 borrowed some museum equipment from the Panzermuseum in Munster. Here such a vehicle, a BMP, can be seen with an Amberland militia member at a checkpoint. (Clemens Niesner)

An Austrian infantryman has taken up position behind a sandbag fortification at a German checkpoint that is being attacked by irregular forces. The soldier is armed with a 7.62mm MG74 machine gun, the Austrian version of the MG3. The weapon has a weight of 11.7kg and a theoretical rate of fire of 1,200 rounds per minute. The weapon has an effective range of 600m when used with the bipod.

The Duro 3 wheeled armored personnel carrier is just entering service with the Bundeswehr. In the German Army the vehicle will see service as a command post vehicle for Luna UAV units, as EOD unit transport, a medical vehicle, and as a vehicle for the Military Police. The 6x6 Duro is powered by a six-cylinder turbocharged intercooler diesel engine that develops 184kw. This engine allows the vehicle with its combat weight of 12 tons to reach a top speed of 100km/h. The pictured vehicle is armed with a 40mm automatic grenade launcher made by Heckler & Koch. (Clemens Niesner)

During their demonstration at the Bergen Hohne ranges, Jägerbrigade 37 was supported by a mechanised infantry company from the Austrian Panzergrenadierbataillon 9. The Austrian company deployed to Germany with their new Ulan armored infantry fighting vehicles. The 28-ton Ulan is armed with a 30mm Mk30-2 stabilised Mauser cannon and has a crew of three, consisting of a commander, driver and gunner. In the rear fighting compartment of the vehicle there is space for eight infantrymen and all their equipment.

Leopards against Leopards
German-Polish Tank Maneuver "Salchauer Schild"

Walter Böhm

A Polish Leopard 2 tank company is identical in structure to a German tank company. There are three tank platoons, each having four main battle tanks, plus one tank for the company commander, and one for the executive officer. This gives a total of fourteen main battle tanks per company.

It is early in the morning. The autumn fog is still lying over the Altmark-Letzlinger Heide. Heard over the radio is the order: "Engines on!" The drivers of 2./PzBtl 203 crank up their tank engines. Rumbling and roaring fills the air in the assembly area of combat unit 203. Large, dark shadows loom over the ancient trees, which have seen the passage of generations of Russian soldiers in bygone times. The combat unit leaves the assembly area in an urgent, yet orderly, manner in the first morning light. Tanks, armored infantry riflemen, combat engineers, artillery observers, armored recovery vehicles and wheeled armored transport vehicles roll eastwards in the direction of the line of departure. Armored infantry riflemen are in front, while behind them are the "huge tanks", the new Leopard 2A6 main battle tanks of PzBtl 203. The accompanying Polish tank company is also on the march. They are tasked with engaging the enemy front lines and keeping them occupied. The mechanized infantry fighting vehicles, Marder 1A3s, of PzGrenBtl 352 drive on ahead to secure the approaches for the main battle tanks. The tension rises as the line of departure is crossed. "Charlie, march! Forward march!" This is the signal to begin the attack. It is the third combat day of the GÜZ 10-04 training event in the Letzlinger Heide during the maneuver "Salchauer Schild". The order today states: "We attack to win back the lost line on the front border of defense (VRV/Vordere Rand der Verteidigung)."

An Historical Event
In the late summer of 2004, history was made at the Letzlinger Heide Training Area. For the first time since the end of WWII, a Polish tank

A break from the battle: a Polish Leopard 2A4 on Heldenweg at the GÜZ. The German-supplied Leopard 2A4 main battle tanks were transferred to the Polish troops as is. Equipment includes leather helmets for the tank crew, German air-defense 7.62mm MG 3s, and German radio sets.

It was a quantum leap for the tank crews of the 10th Polish Armored Cavalry Brigade when they moved from their T-72/PT-91 main battle tanks to German Leopard 2A4 main battle tanks. Training for the tank crews of the 2nd Polish Armored Battalion was provided by the German PzBtl 203 (Hemer).

To destroy a Leopard 2A4 main battle tank frontally in the Gefechtsübungszentrum Heer (Army Combat Training Center), two to three 120mm KE rounds are required.

company, equipped with German 2A4 Leopard main battle tanks, trained alongside a German combat unit. During the GÜZ (Gefechtsübungszentrum Heer) training event 10-04 of Tank Brigade 21 of the 7th German Panzerdivision, the 6th Company of the 2nd Polish Tank Battalion of the 10th Polish Armored Cavalry Brigade was assigned to combat unit 203 (Panzerbataillon 203). This, so far unique, cooperation with a former Warsaw Pact member actually began way back in 2001.

Background

After the fall of the Iron Curtain, a radical change also commenced for the Polish armed forces. Of the former 400,000 soldiers of the Polish land forces, only 90,000 were destined to remain in the targeted armed forces structure of the present day. This structure was to be organized into four divisions, four independent brigades and two air mobile brigades. Within the scope of the NATO alliance, Poland is aiming for the same mechanized brigade structure that is common amongst other NATO partners. To act as the "model brigade", the 10th Polish Armored Cavalry Brigade of the 11th Polish Armored Cavalry Division was selected. The 10th Polish Armored Cavalry Brigade is stationed in Swietoszow, which is located between Cottbus and Wroclaw (Breslau). As the first step of integration, the 10th Polish Armored Cavalry Brigade was placed under the auspices of the German 7th Panzerdivision in Düsseldorf, as part of NATO's reaction force. This essentially means that the 10th Polish Armored Cavalry Brigade is assigned to the German 7th Panzerdivision in the case of an emergency. It was agreed that the future weapons of this Polish Brigade should consist of German military equipment.

The Polish tank company lost a major portion of their Leopard 2A4 main battle tanks through hostile attacks in the first night of the 72-hour combat exercise. But the AGDUS training system simulates repairable damage as well as catastrophic kills, meaning such vehicles can be repaired during combat. This way, the 2nd Polish Tank Battalion was able to salvage 'damaged' tanks and repair them. Therefore, a 100 % state of readiness could be reestablished.

The German tank delivery to the Polish Army included one set of replacement parts for the Leopard 2A4 main battle tank. To carry out the necessary maintenance and the daily handling of the main battle tanks, the 2nd Polish Tank Battalion crews were trained by instructors and technicians of the German 7th PzDiv (Tank Division).

German Materiel and Equipment

In the year 2001, a decision was reached by the Polish Government to not only operate the equipment of a German tank brigade, but to establish the Leopard 2A4 main battle tank as the principal Polish tank. The subsequent German military equipment delivery to Poland included 128 Leopard 2A4 main battle tanks, 4 Biber bridge-laying tanks, 10 Bergepanzer 2 Standard armored recovery tanks, 35 tracked M113A2 (various types) vehicles, 25 Mercedes Benz GD "Wolf" cross-country vehicles, 6 SLT Elefant heavy-duty transporters, and 120 trucks of varying types. As well as these vehicles, replacement parts, tools and training equipment were also delivered. Furthermore, the package included training for Polish personnel within both Germany and Poland, including a logistical support group. At that time, the 10th Polish Armored Cavalry Brigade had a strength of one tank battalion with T-72/PT-91 main battle tanks, an armored infantry rifle battalion with BMP-1 armored personnel carriers, an air-defense battalion, and a tank artillery battalion.

German Training Support

The 14th and 21st Tank Brigades of the 7th German PzDiv were put in charge of training the Polish operational leaders, who had to adapt from their existing T-72/PT-91 main battle tanks to the new weapons systems of the Leopard 2A4 MBT. The 1st Polish Tank Battalion was attended to by the German PzBtl 64 (Wolfhagen) and the 2nd Polish Tank Battalion was trained by PzBtl 203 (Hemer). Here, in contrast to the 1st Polish Tank Battalion that was being trained in its entirety (including crew training in Germany), the "snowball system" was to be applied. This meant to begin with, training Polish leaders in Germany, who would then in turn pass on their knowledge to crews in Poland. From 2001 on, training of Polish leaders began in PzBtl 203, as well as in the tank troop school in Munster, Germany.

The position of the air-defense 7.62mm MG 3 on the turret is typical for the Polish Leopard 2A4 main battle tank of the 6th Company, 2nd Polish Tank Battalion. The Polish Army is discussing modernization of their T-72 main battle tanks based on the Leopard 2 series for improvements in their national interoperability.

A close-up view of the tactical markings on the turret of a Leopard 2A4 main battle tank of the Polish tank battalion. The four-digit number does not give information about the company (as it usually does on German tanks), but is merely used for the consecutive numbering of vehicles. The hussar helmet identifies the 11th Polish Armored Cavalry Division; underneath, the number "2" identifies the 2nd Polish Tank Battalion.

On 16 September 2002, the ceremonial handover and commissioning of the first Polish company with fifteen Leopard 2A4 main battle tanks took place in Swietoszow, Poland. The Leopard 2A4 main battle tanks, supplied by Germany, mostly originated from the disbanded Panzerbataillon 294 and Panzerbataillon 304. Before handing them over, the vehicles were returned to a pristine condition (new track pads, etc.) and provided with Polish vehicle license plates. Otherwise, there were no alterations made to the exteriors or interiors of the Leopard 2A4 main battle tanks. At the end of 2002, the first visit of the 1st Polish Tank Battalion at a training area of the German Panzerbataillon 64 occurred near Swietoszow. Starting in the middle of May 2003, PzBtl 203 conducted training of the 2nd Polish Tank Battalion directly in Poland. Understandably, the language barrier continued to be a problem throughout training. The goal of the training at PzBtl 203 was a combined GÜZ training event with a Polish Leopard 2A4 tank company of 2nd Polish Tank Battalion. After ups and downs, the training stage was reached, and there were no further obstacles standing in the way of a joint German-Polish maneuver at GÜZ. To improve cooperation at the multinational level, PzBtl 203 made the utmost efforts. On the occasion of this joint exercise, though, it was clear that considerable differences still existed in the tactical training and carrying out of orders between both of these NATO partners.

Joint German-Polish Tank Maneuver near Magdeburg

In the new Army Combat Training Center (Gefechtsübungszentrum Heer,

Polish tankers. In contrast to other Polish army units, the members of the armored units wear a black uniform, similar to that of the British Royal Tank Regiment. The black uniform is worn for traditional reasons.

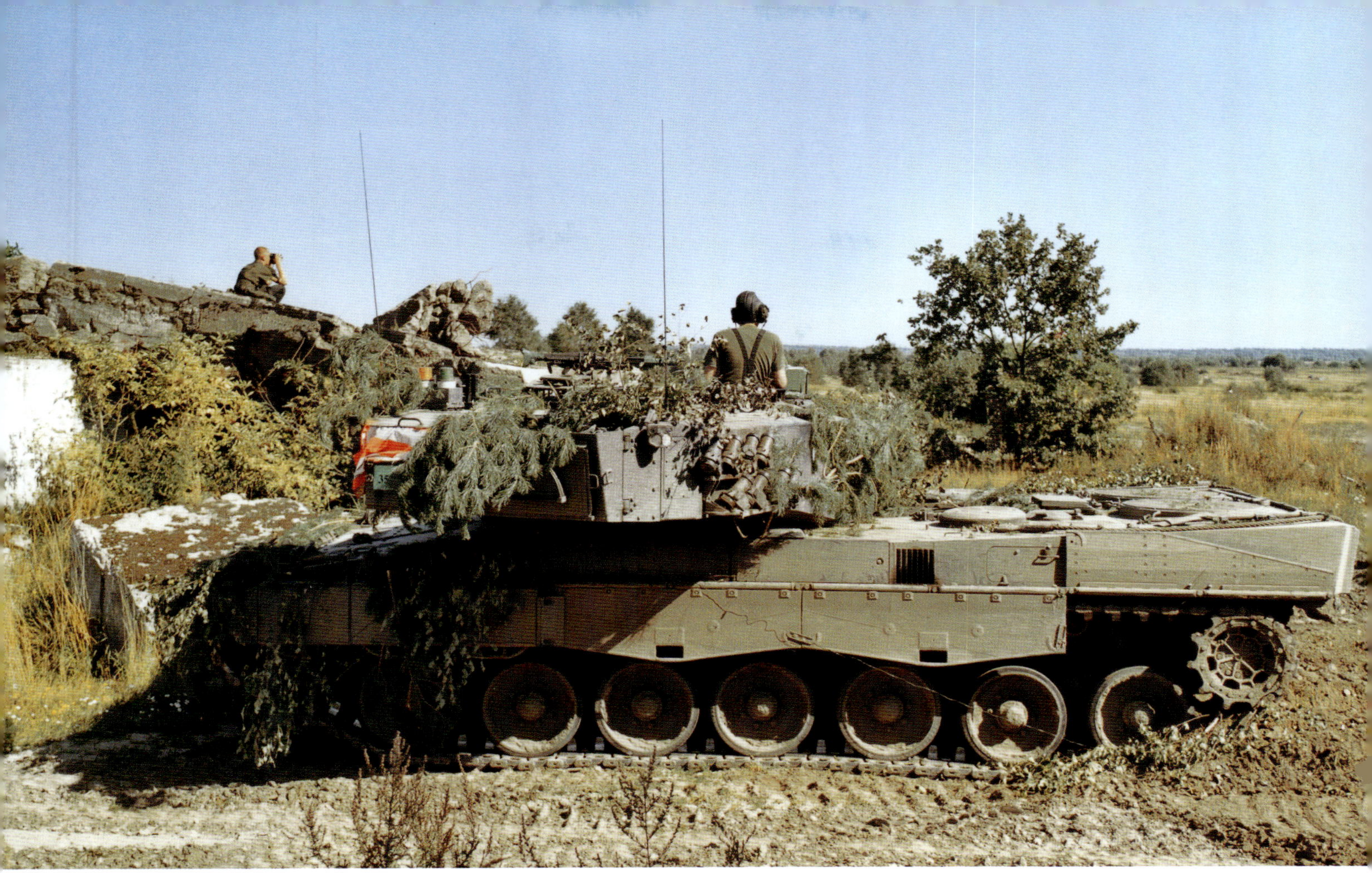

Engagement on the front line of defense (VRV). While the battle tank stays under cover, a dismounted crew member observes "no man's land". This kind of combat leadership was unknown to the Polish unit and was taught by the German instructors during the GÜZ training event.

or GÜZ) of the Bundeswehr in Letzlinger Heide near Magdeburg in eastern Germany, the first German-Polish tank maneuver was to take place to test the lessons learned and to improve capabilities. The goal of the GÜZ training event 10-04 in August-September 2004 was to conduct combat with a highly mobile type of defense. The commander of PzBrig 21 pursued the intention of conducting combined weapons combat training at the reinforced combat troop battalion level. To this end, PzBtl 203 was the lead formation, along with the inclusion of 6th Polish Tank Company of 2nd Polish Tank Battalion. At the end of August 2004, the combat unit transferred to the Army Combat Training Center (GÜZ) near Magdeburg. Due to the combat type of defense, the themes of "fighting for barriers", and "movement in the combat zone", were practiced.

The "Salchauer Schild" maneuver started with a two-day period of platoon training. The Polish Leopard commanders led their tanks safely and competently, although some retraining was needed in fire and movement coordination. The liberties that German platoon leaders have, in accordance with German regulations, were new to the Polish tank platoons. After the progress made in platoon training, the Polish tank company stayed focused, and company training began. With the support of interpreters, the GÜZ instructors did a good job of conducting company training.

During the ensuing battalion training, a non-commissioned officer fluent in both Polish and German was present as a tank cannon loader with the Polish company commander, who was responsible for communications between the German battalion and the Polish company. The battalion training was the basis for the 72-hour combat training that followed, of which the brigade commander of PzBrig 21, was in charge. After two days of technical servicing of equipment and vehicles, combat training began on Sunday, 5 September 2004. The Polish Leopard tank company was engaged on the front border of defense (VRV) between the reinforced 2nd Company of Panzergrenadierbataillon 352 and the reinforced 5th Company of Panzergrenadierbataillon 294. In the organization, these two units do not belong to PzBrig 21, but to Armored Infantry Rifle Brigade 30 of the 10th German Panzerdivision (tank division).

The 7th German Panzerdivision

The 7th Panzerdivision is a large unit of the German Army and in peacetime is placed under the Army command staff in Koblenz. For missions it is integrated into the Allied Command Europe Rapid Reaction Corps (ARRC), under the control of NATO. The ARRC is the multinational Crisis Corps of NATO. Besides Tank Brigade 21 and Tank Brigade 14, in such missions the 10th Polish Armored Cavalry Brigade of

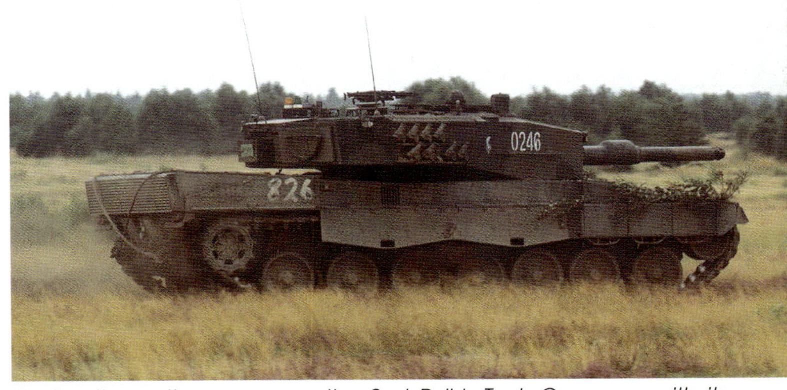

During the entire maneuver, the 2nd Polish Tank Company with its Leopard tanks, remained one unit, and was not divided into platoons.

This Polish Leopard 2A4 moves into position next to the remnants of a former German (Wehrmacht) Army bunker. On the first day of the 72-hour combat exercise, the Polish armored company was placed on the front border of defense and was kept busy defending against enemy reconnaissance forays.

Swietoszow in Silesia also reports to the 7th Panzerdivision. The German units of this division are divided between the Federal States of North Rhine-Westphalia, Hesse, Rhineland Palatinate and Lower Saxony. The main task of the German 7th Panzerdivision is primarily national and alliance defense. The division belongs to the crisis reaction force of the Army and has been involved in both Bosnia Herzegovina and Kosovo. During their missions in the Balkans, the 7th Panzerdivision acquired a great reputation. It also made a considerable contribution to the fight against international terrorism within the scope of "Operation Enduring Freedom", and since the beginning of 2002, has provided troops for the "International Security Assistance Force" (ISAF) in Afghanistan. Under the new structure of the future German Army, the 7th German Panzerdivision will be disbanded.

10th Polish Armored Cavalry Brigade

In the event of a crisis, the 10th Polish Armored Cavalry Brigade will be strategically commanded by the German 7th Panzerdivision. At the leadership level, the 10th Polish Armored Cavalry Brigade is organized with a Staff and Headquarters Battalion, plus it has integral logistics elements. In terms of combat troops, the brigade is provided with two tank battalions (1st and 2nd Tank Battalions) with a total of 116 Leopard 2A4 main battle tanks. The mechanized armored infantry rifle battalion is provided with 58 BMP-1 armored personnel carriers. The reconnaissance company is equipped with BRDM scout vehicles and light motorcycles. An artillery battalion with 18 self-propelled 122mm howitzers (2S1), an air-defense battalion with ZSU 23-4 guns and GROM anti-aircraft rocket launchers, an engineering company (with a German Biber bridge-laying tank), Polish MTLBs, a maintenance company, a support company and a medical company also report to the 10th Polish Armored Cavalry Brigade.

Polish cooperation with NATO has been considerably improved by its taking over German equipment. This photo shows the refueling of a Polish Leopard 2A4 main battle tank by a German MAN 4510 tanker from PzGrenBtl 352, south of the Rabenberg in the GÜZ (Gefechtsübungszentrum Heer).

German GÜZ instructors, Polish interpreters, and Polish Leopard 2 tank commanders conduct a conference. The GÜZ instructors are experienced specialists in tank tactics, who have already done a good job of training the 1st Afghan Armored Battalion (armed with the T-62 MBT). Notice the soldier with the PM 98 machine pistol.

The 10th Armored Cavalry Brigade has participated in various peacekeeping operations of NATO, such as SFOR and KFOR, as well as a stabilizing force mission in Iraq from August 2003 until February 2004.

From CMTC-Hohenfels to the GÜZ, Altmark-Letzlinger Heide

In 1993, the German Bundeswehr started training their mechanized combat troops at the 7th U.S. Army's Combat Maneuver Training Center (CMTC) in Hohenfels. At that time, a new dimension to training began for the armored and associated units of the German Army through utilization of the laser-supported MILES II duel training system, and through umpires in the observer/controller teams (the operations group). The goal of the instruction at CMTC was, besides improvement of leadership training, to train personnel in the use of combined weapons just like occurs on the battlefield. Through the use of the MILES II training system, the soldiers were forced into combat-like behavior, and they learned faster and better than through conventional training. The days where umpires decided about success or failure were over. The experiences and knowledge of the CMTC training events in the '90s served the German Army well, and formed the foundation for their own Army field exercise center.

Army Combat Training Center (GÜZ): "Train as you fight!"

The operational beginning for the world's most modern Army Combat Training Center (GÜZ – short for Gefechtsübungszentrum Heer) of the Bundeswehr in Altmark, near Magdeburg, began in September 1997. It commenced with company training as the first training level, due to an

The 10th Polish Armored Cavalry Brigade is equipped with a variety of different vehicles. German vehicles and Polish vehicles are both in use, like this STAR 266 fuel tanker vehicle, seen refueling an MB 250 GD "Wolf".

The German 250 GD "Wolf" cross-country vehicle, manufactured by Mercedes Benz, is already in use with many NATO armies. In the Polish Army, the "Wolf" partially replaces the Russian UAZ-469B (also called the UAZ-31512).

initiative of the German Heeresamt of 1989/90. The training area at Altmark near the city of Magdeburg, formerly used by Soviet units, was selected because here the armored troops could best train in combined weapons combat tactics. The Altmark-Letzlinger Heide training area was not restricted by ranges and the units could move around freely. In 1998, the German Bundestag (the Lower House of the German Federal Parliament) finally approved the expansion of the GÜZ. Thereafter, expansion to training at the reinforced battalion level gradually took place. Afterwards the utilization of as many as 2,600 soldiers at a time became possible. All events and activities on the combat field are shown and evaluated in a control center for later evaluation and final assessment. In February 2003, the initial goal was reached, and the GÜZ was approved by the Bundesamt for Wehrtechnik (Federal Office of Defense Techniques) and Beschaffung (Procurement), and declared fully functional.

<u>Training Course of Events and Training Goals</u>
The emphasis of training at the GÜZ is on "Leadership Training" for company and battalion commanders. "Leadership Training", simply put, is training, and not training evaluation. Through regular training with their units at the GÜZ under realistic mental and physical conditions, company commanders and leaders are trained in combined weapons combat. The company commander is required to lead his company under combat conditions. Like at CMTC-Hohenfels, the combat units at the GÜZ are mixed and organized like actual combat units. Six months prior to each training event, there is a discussion between the training troop commander and the leader of the GÜZ. In this first coordination meeting, the combat type, training goals and training elements are determined. Thereupon the GÜZ provides training records as well as operational management of the exercise until the time of its completion.

After taking over the equipment and weapons systems of a German armored brigade, the 10th Polish Armored Cavalry Brigade executed a reorganization of its old structure to that of a new NATO brigade. The previous organization featured three tank battalions, an armored infantry rifle battalion, an air-defense battalion, and a tank artillery battalion.

Prior to the exercise there is a reconnaissance of the training area by the training troops and GÜZ personnel. The training troop company commander submits his proposals after the end of the reconnaissance to the leader of the GÜZ for the upcoming combat training. After arrival of the training troops, registration and setup of the AGDUS duel simulators takes place. On the fourth day, the station training of units begins. The actual exercise starts on the fifth day at the GÜZ. The companies leave their bivouac areas and move into the assembly area, where they prepare for their mission. In the training room, the company's combat readiness is about to be put to the test. According to the type of combat - attack or defense - the units start planning their engagements. Thereafter, a three-day combat exercise with daily interim discussions takes place. These intermediate discussions occur with all leaders of the exercise troops, from company commander right down to tank commander. Mistakes are discussed so that they will not be repeated the next day. The company commanders depend on cooperation from the combat support troops throughout, before the exercise ends with the technical servicing of vehicles and specific GÜZ procedures. The removal of exercise equipment takes place and final discussions occur.

<u>GÜZ Equipment</u>
At the GÜZ, tanks obviously do not fire tank shells at each other. The performance and capability of weapons are represented through the

"The small difference". In the foreground there is a Polish M113A2 EFTGEAO KrKw (Krankenwagen), while behind it is a vehicle of Panzerbataillon 203. The German vehicle is a modified type (NDV 1/Nutzungsdauerverlängerung 1) of M113A2 KrKw. Notice the two small cutouts above the tracks under the Red Cross markings.

Along with wheeled vehicles, command post vehicles and bridge-laying tanks, this *Bergepanzer 2A2 Standard (Armored Recovery vehicle)* was also delivered to the Polish Army. The 10th Polish Armored Cavalry Brigade received a total of ten 2A2 Standard Wreckers from Germany. Notice the Polish tactical emblem.

The 2./PzBtl 203 company commander is marking hostile mine barriers on his map. At the GÜZ, the company commander is decidedly dependent on the cooperation of his combat support troops such as engineers and artillery. If he neglects this cooperation, he will fail in combat. Notice the typical "tank overalls" of the German tank crew.

The 2./PzBtl 203 company commander observes the approach of hostile combat reconnaissance vehicles from the turret of his Leopard 2A6 main battle tank. The Leopard 2A6 main battle tank is standing in a fully concealed position near the Stenneckenberg at the Letzlinger Heide Training Area. Take note of the fact that the Leopard still carries the old turret number "500" of 5./PzBtl 203, which has since been disbanded.

training device duel simulator (AGDUS). Both the firing and the effects of firing real ammunition are simulated through lasers and laser echoes, similar to the American duel system known as MILES II. With AGDUS, targets can be fired at, plus the firer can also become a target in return. To prevent being hit by the enemy's AGDUS laser beam, the exercise participants have to behave as if in an actual war. The same kind of realistic behavior is demanded from all, whether an individual rifleman or a tank or howitzer crew. It starts with the correct selection of 120mm ammunition for the Leopard 2 main battle tank's cannon, as well as the amounts of ammunition and fuel that are available for the tanks. AGDUS reference modules are attached to all the vehicles and soldiers involved in the GÜZ training event, which consist of reflectors and laser detectors. On the turret of Leopard 2 main battle tanks, another four modules are attached. In addition, the Leopard 2 is provided with a hull sensor that shows if the tank was in a hull-down, or in-the-open position while firing. There are two reference modules on the front of the turret of the Marder 1A3 tracked armored infantry fighting vehicle. Vehicles with weapons systems are provided with AGDUS Active and vehicles without weapons are provided with AGDUS Passive. The reason for this is to force realistic combat behavior.

According to their missions and participating units, the deployed soldier is provided with AGDUS equipment for personal weapons and body equipment. When a soldier gets hit, a loud beeping tone sounds. To turn off this beep tone, the soldier has to take a red key out of the laser unit of his weapon and insert it into the evaluation unit. Then he drops out of combat and can no longer participate in the exercise. This data is all recorded in the exercise headquarters. For battle tanks that drop out of the battle, the turret is moved to a 6 o'clock position. Through a GPS unit, the position of each vehicle on the combat field can be determined and represented on a large screen in the exercise headquarters. Artillery and mortar fire, as well as the effect of mines and mine barriers, are all realistically simulated. For evaluation during the final conferences, all radio conversations of the exercise's "Blue" troops are recorded. Through fade-ins of a hostile Eloka at the exercise headquarters, the Blue Force's radio conversations and communications can be disturbed.

Structure of "Red" Lead Troop

Similar to the OPFOR (Opposing Force) at the American CMTC-Hohenfels, a professional training troop (enemy) was developed in the Army Field Exercise Center, against which the "Blue" training troops have

A field tactical marking of 2./PzBtl 203, on a Leopard 2A6 main battle tank.

Leading from the front! The PzBtl 203 battalion commander in his Leopard 2A6 main battle tank on the Reichsstrasse. One of his tasks is to coordinate and control combat. His success will depend upon exact reports being received from his subordinate officers. Without control in combat, the worst can happen, which is being killed by friendly fire.

to compete. At the GÜZ, this unit is called "Red" lead troop and is organized into a headquarters company, training area management squadron, two reinforced light infantry companies (vstk Jägerkompanie) one reinforced tank company (vst PzKp) and a reinforced armored infantry company (vst PzGrenKp). The two reinforced light infantry companies (vst Jägerkompanien) are provided with Fuchs 1A4 wheeled armored transport vehicles. The armored company is equipped with Leopard 2A4 main battle tanks, and the armored infantry riflemen are equipped with the Marder 1A3 tracked armored infantry assault vehicle. The fighting vehicles of "Red" lead troop are retained in their original condition but furnished with "red crosses", and are not modified with alteration kits to give them a different appearance, as occurs at the American CMTC-Hohenfels.

Summary

In conclusion, it can be seen that the GÜZ offers the basis for very realistic training, as well as an objective and almost instantaneous training evaluation of the exercise. During exercise discussion sessions, all participants find out how to improve their performance in the future ("sweat saves blood!"), and which of their decisions had the most influence on the course of the exercise. That way, the effectiveness of troop training is considerably improved. Besides the nations already using the Leopard 2, such as the Netherlands Army and the Polish Army, other NATO states are interested in the modern training techniques being used at the GÜZ for training their own troops. The emphasis remains very much on the training of military leaders, and their motto is: "Train as you fight!"

A counterattack is carried out by a Leopard 2A6 MBT of Task Force (GV 203) near the Trockental. The Training Area Letzlinger Heide is also used for training German contingents for SFOR and KFOR missions in the Balkans.

Due to its additional armor, the new Leopard 2A6 main battle tank has less of a heat signature than the Leopard 2A4 main battle tank.

A "German Leopard". The PzBtl 203 possesses both the Leopard 2A4 and the Leopard 2A5/A6 main battle tanks. Take note of the combat camouflage of branches and fabric strips.

Here is Kwakye Yeboam. As in many other European armies, appearances get more "colorful" through immigration from more southerly countries. Kwakye Yeboam is the driver of a Leopard 2A4 main battle tank in PzBtl 203 and is excited about his job, which allows him to move through the countryside with 1,500 horsepower at his fingertips.

In German tactics during combat with combined weapons, the tank troop is the priority weapons system. A coordinated "play" between fire and movement is required to bring the full tactical efficiency and fighting strength of the tanks into effect. This type of leadership is standard with German tank troops.

"Look out! Make way!" A Leopard 2A6 main battle tank moves into a rear position in full reverse after its position was discovered by the enemy. Another training goal during the GÜZ training event 10-04 was to practice the principles of disengaging from contact with the enemy in mobility-led defense combat. Rapid backward movement is made easier by a special reversing camera at the rear of the Leopard 2A6.

All thirteen German armored battalions (as of 2004) are provided with the efficient Armored Recovery Tank 3 Büffel. Based on the Leopard 2 main battle tank chassis, the Büffel (with a combat weight of 54.5 tons) was introduced at the beginning of the '90s to replace the lighter wrecker 2A2 Standard (with a combat weight of just 40.6 tons).

Waiting for fuel south of the Rabenberg at the training area Altmark-Letzlinger Heide. In the foreground is a German PzBtl 203 Leopard 2A5, and behind it a 2nd Polish Armored Battalion Leopard 2A4. The GÜZ training event also improved interoperability between the German and Polish Armies.

Not a soldier of the future, but a member of the Austrian Armored Infantry Rifle Battalion 35 with full AGDUS training equipment. The soldier is armed with a G 36 German 5.56mm rifle. Note the soldier's reference (sensory) modules on his helmet and arm.

Setting up barricades. Combat engineers of 1st Armored Engineer Battalion (Holzminden) are setting up a DM-12 anti-tank mine trap in the area of the Trockental. The DM-12 anti-tank mine will penetrate the hulls of most modern battle tanks if it is detonated in close proximity.

Mount! The Austrian armored infantry rifle platoon was assigned to the German 2./PzGrenBtl 352. One of the Austrian armored infantry riflemen secures the mounting of his comrades into a German Marder 1A3 tracked armored infantry assault vehicle during combat with "Red" troops at the Reichsstrasse.

To protect Gefechtsverband 203 from air assault, Roland 1 armored air-defense vehicles of PzFlaRak 300 from Fuldatal were used. Aircraft flying at an altitude of up to 4,000m, and at a combat distance of up to 6,000m, can be engaged with the Roland 1. The Siemens pulse-Doppler surveillance radar is operational in this picture.

An M113GA2 NDV2 Geräteträger-Rechenverbund (GerTrgRechnvbu ADLER) from Panzerartilleriebataillon 215 of 21st Armored Brigade. The ADLER computer system (Artillerie-Daten-Lage- und Einsatz-Rechenverbund) is placed on the rear roof. Note that the placement of the tactical sign on the vehicle´s left front side, over the track, is not typical for German M113s.

This Marder 1A3 belongs to 2./PzGrenBtl 352 from Mellrichstadt. During the GÜZ 10-04 training event, the "Blue" training troops, Task Force (GV 203) consisted of two armored companies and two armored infantry riflemen companies, as well as further combat support troops such as engineers, air-defense and artillery units.

A Dachs 2A1 engineer tank of 1st Armored Engineer Battalion during a change of combat position. With its rotating telescopic arm and large excavator shovel, the Dachs can excavate positions for tanks, and remove barricades or obstacles. Even the best-equipped armies in the world do not have such a vehicle as the German Dachs in their armored engineer companies.

A Fuchs 1A4 wheeled armored transport vehicle of 21st Armored Brigade of Augustdorf in combat camouflage, is on the move. In an effort to reduce dust, fabric camouflage strips are hung over the wheels.

"Idefix destroyed!" as portrayed by the position of its turret at 6 o'clock! The crew of this "dead" Leopard 2A4 main battle tank of the "Red" lead troop obviously liked the dog "Idefix" belonging to big Obelix, from the famous comic book, "Asterix".

"Field outpost at Double H". A Leopard 2A4 main battle tank of the "Red" Force leads troops into position at "Double H". The steel scaffolding of the so-called "Double H" originates from the time when the German (Wehrmacht) armed forces used the Letzlinger Heide Training Area for firing tests, e.g. with the heavy rail "Dora" 80cm cannon. After that time, Russian troops trained in this area. Before its use by the Bundeswehr, the area had to be cleared of tons of old shells, duds and ammunition parts left behind by Russian troops and the former German Wehrmacht. This onerous task took years.

The "Red" lead troop lost another Marder 1A3 tracked armored infantry assault vehicle through artillery fire. Artillery fire is simulated with smoke grenades.

A member of a light infantry company (Jägerkompanie) of the "Red" lead troop moves into an ambush position with his 7.62mm MG 3. All soldiers and combat vehicles participating in the exercise are equipped with hit-sensing receptors.

This Leopard 2A4 main battle tank of the "Red" lead troop (distinguished by red crosses) carries out combat reconnaissance in a Polish armored company defense sector. The goal of the combat reconnaissance is to reconnoiter and report the strength, positions, and defenses of the enemy. The "Red" lead troop at the GÜZ has, like the U.S. OPFOR at CMTC-Hohenfels, the distinct advantage of knowing the Area of Operations (AOR) very well.

Besides a tank company and an armored infantry rifle company, the "Red" lead troop has two reinforced light infantry companies (Jägerkompanien), which are equipped with the Fuchs 1A4 armored urban assault vehicle.

BDMZ-III
The Mechanized Landing Battalion of the Spanish Marines

Juan Pablo Lasterra

Two Marines provide cover for their comrades during a training session in the CASR. The main light weapons of the Spanish Marines are the H&K G36E assault rifle, the Ameli light machinegun (both have a 5.56mm calibre), the Accuracy International AW and Barret M95 sniper rifles (7.62mm and .50-cal. respectively), and the Instalaza C90 90mm rocket-launcher.

In early 2004, the Spanish Brigada de Infanteria de Marina (BRIMAR) received eighteen Mowag Piranha III 8x8 armored amphibious vehicles. These vehicles were purchased to equip the two mechanized companies of the Batallón de Desembarco Mecanizado (BDMZ-III). The vehicles were acquired under a €45.4 million contract that was signed in 2001. This purchase was made within an even more comprehensive program that provides the option of obtaining a total of thirty-eight armored vehicles for the Marines in three phases. Integrated logistics support, driver training, training of maintenance personnel (including refresher courses), and technical documentation, were included in the initial contract.

The eighteen Piranhas will serve to increase the BRIMAR's amphibious assault capacity and to expand the options available during intervention in peacekeeping and peace-enforcing missions. The

BRIMAR participates in Multinational Forces like the European Marine Force (EUROMARFOR), the Spanish-Italian Amphibious Force (SIAF) and the Combined Amphibious Force of the Mediterranean (CAFMED).

The BRIMAR's BDMZ III had its origins in the Grupo Anfibio Mecanizado GAM (Mechanized Amphibious Group), disbanded in 2001. It is composed of a Command and Services Company, two Infantry Mechanized Companies (9ª and 10ª) and a Combat Tank Company (11ª). This battalion has the standard missions of an infantry battalion, but with the added characteristics of speed, protection and firepower as provided by Piranhas and M60A3 tanks.

The eighteen BRIMAR Piranha IIIs have been the first success for Mowag in Spain. The Spanish Marines' modernization program contemplates obtaining a total of thirty-eight armored vehicles. The Piranhas arrived in two lots - at the end of 2003 and in March 2004. First to arrive were the APC versions, followed by the Section, Company and Battalion Command Posts and the Ambulance.

A lieutenant instructs the platoon commanders that are soon to take part in an exercise in the CASR, the field training camp of the Spanish Marines.

Each of the Mechanized Infantry Companies is divided into two sections, both of which are commanded by a lieutenant, with four Piranhas per section. A section is composed of three infantry platoons and a weapons platoon. Each Piranha carries a platoon of eleven Marines: the driver, gunner, and commander, plus two squads of four riflemen. The 11ª Combat Tank Company is divided into three sections, each one equipped with four M60A3 combat tanks, and commanded by a lieutenant.

THE SPANISH PIRANHAS IN DETAIL

The Piranha III belongs to a well-known family of vehicles developed by the Swiss company Mowag Motorwagenfabriken, nowadays part of General Dynamics Land Systems-Europe (GDLS-E). The third generation of Piranhas was developed in 1996 and is based on a structure made up of electrically-welded, hardened steel armor plates that provide increased ballistic protection against direct hits from automatic weapons and artillery shell splinters. An additional fast-assembly armored kit can be fitted to improve the overall protection level of the vehicle.

Another advantage of the III series is its higher yield in terms of payload. Its usable internal space for weapons stations, equipment and troops is greater than in the first models. Other characteristics include: a power pack on the front right-hand side, central driveline system, independent suspension, built-in fire suppression system, and ample space for the stowage of arms, communications gear and troops. A C-130 Hercules can transport a Piranha III completely equipped and ready for immediate deployment.

A platoon of Marines mounting their Piranha. Each vehicle carries eight riflemen divided into two squads, along with the squad commander, gunner and driver.

This first sale of Mowag vehicles to Spain has brought about the arrival of the following vehicles in the BRIMAR:
- A Battalion Command Post without a turret. It is equipped with a Krauss-Maffei Wegmann (KMW) assembly mounting an M2 HB .50-cal machine gun, an AN/TVS-5 night vision device, four double M-246 66mm smoke grenade launchers, PR4-G communications equipment, an extendable command post tent, and an auxiliary power unit.
- A Company Command Post provided with a rotating single-seat Cadillac Gage-Textron turret. This variant is equipped with an M2 HB .50-cal machine gun, a Mk.19 Mod.3 40mm automatic grenade launcher, eight double M-246 66mm smoke grenade launchers, an M-36E1 periscopic day/night sight with laser aim telemetry system, and seven vision blocks. This turret is therefore an updated variant of the one mounted in the AAV-7 amphibious assault vehicles of the BRIMAR, and similar to those in service with the U.S. Marine Corps under the Up-gunned Weapons Station (UWS) designation.
- Three Section Command Posts.
- Twelve Armored Personnel Carrier (APC) vehicles for the platoons.
- A Piranha III Ambulance fitted with four stretchers and advanced medical equipment.

In this picture we can appreciate the external appearance of the vehicle from the rear. The CASR camp offers a wide variety of terrain for the drivers to train on in their new Piranhas.

Frontal view of a Spanish Piranha III. Note the national flag. €45.4 million was the total purchase price for the eighteen vehicles.

A Piranha III crosses over the path of a Humvee. The Spanish Marines are proud of having the best military vehicles available for each task.

The Piranha III silhouette is destined to be a classic image of the modern Spanish BRIMAR, as the M60A3TTS, AAV-7 or Alvis Scorpion have been in the past.

All the Section Command Posts and APC vehicles are externally identical to the Company Command Post. The only difference lies in the internal radio transmission equipment that varies depending on which command echelon the vehicle is serving in.

The BRIMAR's Piranha IIIs have some characteristics that are taken from the series IV. The power unit is a 400hp diesel Caterpillar C9 (3126) engine with 6 cylinders in line, and with a ZF Ecomat-7/HP 602 automatic transmission. This power unit also feeds the two propellers positioned at the rear for water propulsion. The amphibious kit comes complete with a protected gas escape system and a folding front breakwater screen. The manufacturer made these modifications in order to fulfill the Spanish requirements for amphibious combat.

Each wheel has independent suspension and hydraulic damping, with front shock-absorbing coil springs and rear torsion bars. These produce a favorable all-terrain quality for the vehicle. A Central Tire Inflation System (CTIS), controlled by the driver, regulates the pressure of the tires. In an amphibious assault operation, it is advisable to have low pressure in the tires, with the purpose of obtaining a greater traction surface when in sand. The brake system is a dual circuit, air actuated ABS 6S/C.

The Spanish Piranhas are equipped with a monitoring video camcorder and a laser warning system. The camera is the British-manufactured Ultrak system, and it provides the platoon's second-in-command with images of the perimeter of the vehicle. Located on the upper rear section of the Piranha, the Ultrak is capable of rotating horizontally and vertically providing a wide field of vision. For that reason, this equipment would be very useful during peacekeeping operations, or when the vehicle is performing complicated maneuvers. Images from the video system are displayed on a retractable screen located in the interior of the Piranha, next to the rear access ramp.

The laser warning system, consisting of four sensors located on each side of the vehicle's roof, is an Avimo-Thales Optronics LWD2. Each sensor's field of view is greater than 180°. If the Piranha is illuminated by a hostile laser device, the signal is detected by the corresponding sensor which sends the information to an interior control/display panel via a junction box. Visual and audible signals warn the crew of impending danger. The angle from which the threat is coming is reflected on the warning system's panel, allowing the crew to take evasive action.

An overpressure system has been chosen as a NBC protection kit. Although it protects the occupants while they remain in the Piranha, the Marines have to be dressed in their NBC suits before opening any of the hatches or the rear ramp. In addition to all the elements already mentioned, the Piranha has a GPS and a compass system to aid in navigation.

MANEUVERING IN THE RETÍN

Located near the southern Spanish town of Zahara de los Atunes, the Campo de Adiestramiento de la Sierra del Retín (CASR) serves as the field training camp for the diverse units of the BRIMAR and other allied units. The CASR is where we were invited by Lieutenant Colonel Melendez Pasquin, commander of the BDMZ-III, to see the recently acquired Piranhas in action.

A platoon of Marines disembarks from their Piranha III. The AAV-7 will remain the main amphibious assault vehicle of the BRIMAR, with the Piranhas being more suited for peacekeeping operations of the Infantería de Marina.

After a simulated assault, a platoon of Marines approaches its vehicle in order to re-embark. Note that a Marine provides security until all the riflemen have boarded the vehicle. Additional cover is provided by the Piranha gunner.

The last members of a squad embark inside a Piranha III. Note the C-90 grenade launcher. The Piranha is armed with an M2 HB machine gun and a Mk.19 automatic grenade launcher. The possible arrival in the future of a reconnaissance Piranha armed with a 40mm cannon will increase the firepower of the BDMZ-III.

The last Marine who was providing security for his comrades embarks on a Piranha III using the rear ramp door.

A platoon of Marines sitting inside the Piranha. As part of the overall package, a number of systems have been installed in the vehicle.

The crew of a Piranha observes the advance of their disembarked squad. The Spanish Piranha III APCs are equipped with a Cadillac Gage-Textron turret, armed with an M2 HB .50-cal machine gun and a Mk.19 40mm automatic grenade launcher, identical to that used by the U.S. Marine Corps under the designation of the Up-gunned Weapons Station (UWS).

We travel to a wooded area where two platoons of Marines rest inside their individual igloo-shaped tents. A lieutenant issues the embarking order and the Marines pick up their weapons and quickly load into the Piranhas, parked next to the encampment. When everything is ready, both vehicles begin to move across the rough terrain, allowing us to immediately verify the high maneuverability of the vehicle.

From an open area we can see the Piranhas moving across a field at a high rate of speed. Arriving at a previously determined point, both vehicles stop at the same time and four squads of riflemen disembark through the rear ramp. They begin the assault of an imaginary enemy position, while the platoon commanders observe the assault from inside their vehicles. The gunners also observe the action from their turrets, ready to provide support fire with their M2 HB machineguns and Mk.19 grenade launchers if necessary. Once the assault is completed, the squads climb back into the Piranhas before leaving the zone at the same high speed with which they arrived. This same operational procedure is repeated several times, allowing us to take different images of the vehicle and troops in action. During all these repetitions, the run times are similar, which indicates a fast and positive synergy between crews, embarked troops and vehicle.

Later in the morning we have the opportunity to see some M60A3 TTS tanks belonging to the Battle Tank Company of the Battalion in action. Some of the tanks are painted, like the Humvees and Piranhas, with the new camouflage colors destined for use on all the BRIMAR's vehicles. Some vehicles still retain their older patterns. All the M60A3 TTS tanks are currently undergoing a modernization process in which their engines, transmissions, 105mm gun barrels, turrets and sight systems are being improved. The Spanish Marines decided to keep the M60A3 TTS and not buy Leopards (the main battle tank of the Ejercito) due to the heavier weight and larger dimensions of the German tank.

The nineteen AAV-7A1 amphibious assault vehicles and the seventeen Alvis Scorpion reconnaissance vehicles, both part of the old Mechanized Amphibious Group, are not included in the new battalion. The AAV-7A1s, along with their crews, have been transferred to the Grupo de Armas Especiales (GRAE) of the BRIMAR. They are still the primary amphibious assault option for the Spanish Marines and will remain in service until 2010. The Scorpions have been retired from duty due to their obsolescence after many years of excellent service. Since a Mechanized Reconnaissance Section will be created in the future within the BDMZ-III, the most probable option is that the BRIMAR will choose a

In the CASR training camp we were able to see the fast and positive synergy between crews, embarked troops, and vehicles.

reconnaissance Piranha model to replace them, equipped with a 40mm cannon. However, a decision has not been made on this matter yet.

CONCLUSION

The new Piranha III vehicles are destined to extend the BRIMAR's amphibious assault capacity and to improve the BRIMAR's performance in peacekeeping missions. With all the modifications mentioned above, the first goal will be fulfilled without any problems. With respect to the second aim, the vehicle's armament, ballistic protection and built-in security elements have been thought through specifically for this type of mission. The future arrival of a better-armed reconnaissance Piranha will further increase the range of possible missions for the BDMZ-III.

Close-up photo of one of the two propellers located in the rear for propulsion in water. The maximum speed in water is five knots and the vehicle can maneuver for four hours in that element.

The versatile Ultrak British-made camera provides the crew with video images of the exterior. This can be a useful tool in peacekeeping missions.

A Marine observes the retractable screen in which the images sent by the Ultrak camera, positioned above, are displayed.

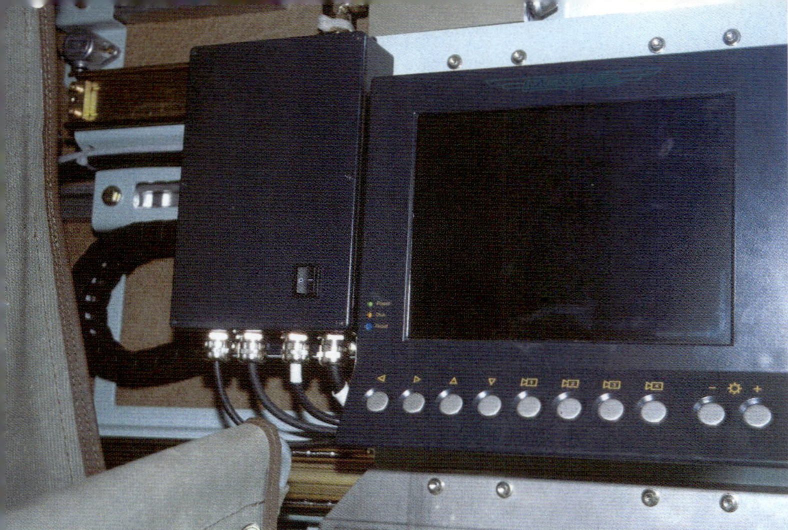

This is the retractable screen, located next to the rear access ramp door, on which the video images gathered by the Ultrak camera are displayed. The second commander of the squad is charged with the responsibility of operating the video monitoring system.

Close-up shot of one of the four sensors of the Avimo-Thales Optronics LWD2 laser warning system.

One of the control panels located in the interior. Note that all the instruction labels are written in Spanish.

TECHNICAL DATA FOR THE SPANISH PIRANHA III 8x8

Crew:	3 + 8 (driver, machine gunner, commander and 8 riflemen)
Weight empty:	12.5 tons
Combat weight:	18.5 tons
Payload:	6 tons
Length:	6.93m
Width:	2.66m
Height:	2.17m
Power to weight ratio:	21.62hp/ton
Voltage:	24V
Internal volume:	11m^3
Armament:	1x M2 HB .50-cal machine gun and 1x Mk.19 40mm automatic grenade launcher.
Gun sight system:	M-36E1 day/night
Suspension:	hydraulic; independent for each wheel
Maximum speed:	105km/h on roads and 5 knots in water
Operational range:	480km on roads and 4 hours in water
Gradient:	60%
Trench crossing ability:	2m
Engine:	Caterpillar C9 (3126) 6 cylinder diesel developing 400hp
Transmission:	automatic ZF Ecomat-7/HP 602 with 6 forward + 1 reverse gears
Communications systems:	PR4G and Rovis
Other features:	dual-circuit air-actuated ABS brake system, CTIS system for tires, 2 propellers for water propulsion, air-conditioning system, NBC overpressure protection kit, fire and explosion suppression system, monitoring video-camera, and laser warning system.

The Piranha III weapon station with the gunner's helmet, the .50-cal M2 HB machine gun and the spent cartridge hose.

The second batch of Piranhas, received in March 2004, included this Ambulance. The ambulance variant is fitted with four stretchers and advanced medical equipment. The vehicles were parked at the Spanish Marines' San Fernando headquarters, near the southern city of Cadiz.

Marines are the elite forces of many countries. Spain is no exception and the BRIMAR is an excellent tool of the national armed forces. The 5.56mm cal. H&K G36E assault rifle is the main light weapon of the Spanish Marines. In the near future, the G36KE, a shorter variant of the standard model, will be adopted by some Special Forces and reconnaissance units.

A detailed picture of the weapon station controls. The big monocular is part of the M-36E1 periscopic day/night laser sight system.

47

A lieutenant gives some orders to the crew of an M60A3 TTS tank before an exercise in the CASR. The large size of this training camp allows live fire practice for the BRIMAR's artillery and tanks.

This M60A3 TTS tank shows the current camouflage pattern common to all BRIMAR vehicles. In our visit we were still able to see some tanks of the 11th Combat Tank Company with the old pattern.

Front view of an advancing M60A3 TTS tank belonging to 11th Combat Tank Company of the BDMZ-III. The company is divided into three sections, each equipped with four M60A3 TTS tanks. All the BRIMAR's tanks are currently undergoing a modernization process in which their motor, transmission, gun barrel, turret and sight system are being improved.